LAB LEAK

THE UNTOLD TRUTHS OF COVID-19

NIKUNJ RATHOD

ISBN 979-888546431-4

This book is dedicated to those people who suffered unbearable pain and died due to this dreadful epidemic.

Dedicated to the brave frontline heroes involving Doctors, Medical and Pharma Professionals, Police Administration, and Scavengers who religiously did their duty regardless considering the risk involved for Self and their own family during the COVID-19 pandemic.

The 50% of the profit from the sales of this book will be donated to help the children who lost their parents during the COVID-19 pandemic.

Also

Dedicated to our country's Armed Forces for their services to our Nation.

Contents

FOREWORD

"The books that the world calls controversial and immoral
are books that show the world its own shame." – Oscar Wilde

There is nothing wrong with chaos until it becomes orderly - Zurich axiom.

In the past one and a half years, the COVID- 19 pandemic has given the world its greatest heart attack and pandemonium. The virus spread in a compound fashion and became viral on social media with a lethal bio velocity. Little things, at times, can have infinite consequences. The free-flowing gossip and motivation of the pre-Covid era turned lean on social media. They soon became anxious, and a mental stampede of panic and horror escalated to an outbreak with over 75 a million confirmed cases and 1.7 million dead.

Each passing day made the human species realize how fragile and uncertain life can be. The pandemic has reset the world. Words like 'quarantine' became part of the neo lexicon terminology. The Covid saga is still in vogue, and the hangover seems unending. Today the world knows what ZOOM is. My hands consumed more alcohol from hand wash than my mouth ever did. Also, others empathized with me - how it feels to breathe through a mask. Social media became a medical bulletin, and every third person was a medical adviser.

Once the world regained a bit from this devastating shock, it started to look for how and why it happened. The world, too shocked and suspicious to make an early recovery, still looks for clues. Today, the virus that started its travel itinerary from the Wuhang Lab of China has rambled all over the globe. Covid became as dreaded a word as a nuclear bomb. This epidemic's maverick and savage ways have engaged considerable attention amongst scientists and nations. A controversial nation with its burning ambition to be a supreme power, China made the world suspicious. But, unfortunately, suspicion does not cook legal rice.

Was it an accidental or underhand exercise in intentional biological warfare? The world has seen a nuclear attack; the world saw what biowarfare could do for the first time. Is the pandemic being manipulated and orchestrated by a deep state?. The world was never so helpless and on oxygen. Nothing has confused the world as this mystery microbe. The corona warriors fought valiantly- gave their lives. No one knows exactly knows what happened. God is everywhere, but who is he and where? The pandemic has changed the line and length of life and produced a new version of human beings.

The author lost a few of his loved ones to this grave malady and vowed to bell the cat- and those responsible for it. This research-based book explores the deeper intricacies and tries to find the epicenter and scan the nation behind the COVID epidemic. Bill Gates has been named. Pharma company has been named. Is the virus natural or man-made? Is the vaccine safe or not. The author goes through several research papers, articles, and videos during this journey. Finally, after a great struggle, he reaches- to the smoking gun!

The Covid conspiracy has opened up Pandora's box of no mean proportions. This story is a journey of an author searching for the truth about the emergence of COVID-19. I wish the Author Mr. Nikunj A Rathod, and his revolutionary book '**Lab Leak**' the intrigue, applause, and success it deserves.

Dr. Kamal Murdia,
MBBS, MS, MCh (Plastic Surgery), CS(UK)
Plastic -Cosmetic Surgeon - Author
Celebrated International Speaker
Mumbai, India and Abroad
International Best Seller Author of - The
Million Dollar Powerful Personality and
The Strange Case of the Billion Dollar Tortoise

PREFACE

In 2020, a deadly novel virus suddenly came to the surface. In India, I remember on the evening of 24th March, Prime minister Shri Narendra Modi declared Nationwide lockdown for 21 days, limiting the movement of the entire 1.38 billion population. The whole world was facing an invisible enemy which affected each and everyone's life sooner or later in one or another way. Millions of people succumbed to this invisible enemy, and Millions survived from it after a great fight against it in hospitals.

I remember the exact date I was infected with COVID-19, 6ton. I remember each moment of those 22 days of home quarantine. I preferred to stay in home quarantine instead of going to the hospital. If I had been admitted to the hospital, the morale of my family, my wife, and two sons would have been destroyed completely. I remember begging for Death due to intolerable pain in my throat and chest. Still, my family, especially my wife, kept me motivated and kept giving me the courage to fight with this invisible enemy. Finally, after a passionate fight after 22 days, my RT-PCR report came negative. But the damage was done by that invisible enemy was huge. My stamina had been decreased to a considerable level. However, with the help of my family, it recovered in 2-3 months.

At the end of the year 2020, the effect of the corona virus has diminished. People were hopeful for normal life in the year 2021. But on the contrary, the impact of Covid-19 increased back to the ghat, resulting in a second wave of coronavirus from the January-2021 end, which progressed to its peak by April-2021. This time the virus was showing its most frightening form. This time its symptoms were completely different from the mutations of the previous Covid-19 virus; in many cases, the symptoms were normal or non-existent. Now this invisible enemy had become even more dangerous.

I do not write books for a living, but During this time, what I felt and my curiosity about Covid-19 gave birth to this book.

My only motive behind writing this book was that the truth of the origin of Covid-19 should reach every common human being, which everyone has the right to know. There is absolutely nothing to defame anyone or question anyone's ability behind writing this book. The money earned through this book will be spent on children who have lost both their mother and father during this covid pandemic. Unfortunately, we will never be able to compensate for their pain, but we can help as much as we can.

My only request is to read this book cover to cover.

"Truth Must Be Told"

Best Wishes,

Nikunj Ashwin Rathod, M.Sc. Organic Chemistry
Author- Engineer, Gujarat, India.
E-mail: nikkunj18@gmail.com
Twitter: @authornickunj
Website: www.niekkunjrathod.com

Abbreviations And Acronyms

ARDS - acute respiratory distress syndrome

ARI - acute respiratory infection

BSL- biosafety level

BtSL-CoV – Bat SARS-Like Corona Virus

FAO- Food and Agriculture Organization of the United Nations

HPAI- highly pathogenic avian influenza

ILI- influenza-like illness

IHR- International Health Regulations

IPC- infection prevention and control

MERS-CoV- Middle East respiratory syndrome coronavirus

OIE- World Organisation for Animal Health

PPE- personal protective equipment

PLA – People's liberation army

RT-PCR- reverse transcription-polymerase chain reaction

SARI- severe acute respiratory infection

SARS- severe acute respiratory syndrome

TIPRA- tool for influenza pandemic risk assessment

WHO- World Health Organization

WIV- Wuhan Institute of Virology

Wuhan CDC – Wuhan Center for Disease Control and Prevention

I

The Dilemma

26/06/2021, Saturday

Jamnagar, Gujarat, India

It was a normal day. I woke up at 7:00 a.m. My phone was vibrating with the flow of notifications. I was brushing my teeth and sorting out the important and useless notifications. One of the notifications caught my attention.'COVID cases in last 24 hours: New cases: 48,698. Deaths: 1183'. It was a 'New Normal.' Since December 2019, the definition of the normal world has changed. Normal life means S.M.S (Sanitizer, Mask, and Social distance). I put my phone aside and completed the bathroom routine. I went to the hall and completed my 15 minutes workout. I took a break between bath and workout.

Meanwhile, a hustle was going on in my mind. The deception was about to celebrate or start the new project without any break for celebration. On 24th June, I completed my second book,'**9 secrets of ultimate success,** and submitted the manuscript for further editing and cover design. The hustle was to celebrate the success of my first book '**Fat 2 Fit: The simple science of ultimate building body**', become a bestseller and complete the second book, or write the book I was desperately willing to write. When I was writing my first book 'Fat 2 Fit', I had not expected that the book would become a bestseller in five different countries and more than ten categories. Nevertheless, I was pleased about the book's success and wanted to celebrate it. My heart was now telling me, 'Now is the time to write this book, which you are holding and reading right now.

The dilemma ended soon. Another reason not to celebrate the success was my family. I was working in 'Island' mode due to the COVID pandemic. Many like us have to stay at the company facility without our families. It was already two and half months since I had seen my family face to face apart from video calls. The pictures of the COVID pandemic were unforgettable. The pandemic changed the whole lives of millions of families around the world. Millions of bread earners for the family lost their lives during the pandemic. Millions of children lost their parents during the COVID. A video of a young mother listening to music and dancing to the Bollywood song *'love you Zindagi..'* (Love you life...)fighting against CORONA in hospital on oxygen support vent viral during the second wave of CORONA in India. Billions of people prayed for her around the world.

13th May 2021,

I was sipping my green tea and checking my Twitter account in the morning. I read a condolence tweet from Dr. Monika Langeh, who treated that brave 30-year-old young mother. She informed to media that, "Though doctors tried their best, it was not enough to save her. She was a courageous woman. I have never seen a patient like her." That news broke my heart completely. The woman was the same age as my wife, and the information scared me a lot. Her kid was waiting for her at home, hoping that his mom would come one day, would carry and hug him tightly, but it never happened. That innocent does not become aware that he could not see his mother's face last time. That was even more heartbreaking. This story was of one family. There were millions of families like this. About a threes month before this unfortunate event, another life-changing series of events started in India, which we will uncover one by one further.

12th April 2021,

I got a call from my home. My brother has been sick for the past four days. On **11th April 2021,** he went for the RTPCR testing for CORONA. I picked up the call. It was my father on the other side. He informed me that Darshan, my brother, has tested positive for COVID. My heart sinks for a while. There was fear and worry in his voice. I empathize with him. Put the phone aside. I was in deep fear about my brother. He is not only my brother but also my best friend. On the same day, I got a call from my boss. He told me that you have to quarantine in the 15th April batch. I paused for a few seconds and replied, ' yes, I will.' My wife was waiting to know that whose call was. I explained to her. Her face became dull. It was only nine days that I had been at home. I had taken Emergency leave for my sons.

15th April 2021,

It was one of the most stressful days of my life probably. Nency, my wife, was preparing breakfast. After completing break breakfast, my elder son left for his private coaching. Unfortunately, due to the pandemic, the education system had severely collapsed. So, my wife and I decided to arrange personal coaching for my son. My toddler Vedant was fond of music, so he enjoyed watching songs on T.V. Nency brought breakfast in the drawing-room and served it. She smiled at me. I knew inside she was not even 1% happy. Me, too, feeling angry inside due to the dictatorship-type job system established by the company management. I have never seen this much dumbest management in my ten years of career. I was struggling to start the conversation with her.

She broke the silence and asked me, "Is it necessary to go?"

"I wish I could have denied!" I answered hopelessly.

"How much time will you stay in the township?" She asked, sipping her tea.

"I Do not know, and it might be 3-4 months minimum," I answered.

I changed the T.V. channel to see the news. Vedant turned his face to me and showed his resentment by puffing up his face. He was looking so cute. The news channels were scary those days. Media sells on negativity.COVID cases were rising exponentially. I was apprehensive about my family. I kept thinking about, In my absence, what will happen to them. Nency is a strong lady, but it was challenging to handle two children in such a situation.

15th April 2021, 15:00 Hrs.

I completed a short nap. Vedant was sleeping beside me. He was looking adorable. I wished to kiss him, but I did not want to disturb his precious sleep. Sleep is a crucial element for growing babies. I told Nency to make some tea for me. Dhairya was watching cartoons. He saw me and then to my packed bags. He was aware of my quarantine. Nency served me the tea.

Meanwhile, she made sure that she put all my necessities in the bags. I hugged Nency. As soon as I went to hug Dhairya, he hugged me tightly and started crying. My heart sank, I also started crying. I thought to drop the idea to quarantine for a second, but I was helpless. I hardly found myself this much vulnerable in life. Even though the pandemic was at its peak, there were no rules for quarantine except for international traveling and travelers. After the first wave of the deadly pandemic, the government has finished the law of 14 days quarantine. All employees of the company were suffering from this Hostile Work Environment. The top management people have created such, environment for their benefit. Nency sympathized with me and convinced Dhairya that Dad would come back soon. At that moment, I realized Nency is so brave. It was not easy for her too.

Nency brought the keys of the two-wheeler. I hugged her and left Vedant sleeping. That was not the first time. It happened many times during the pandemic; I saluted the soldiers and their families in my mind. I realized how big a sacrifice our brave soldiers make for our country. I was experiencing that.

Nency dropped me at the quarantine center. She hugged me and left. I kept watching her until she disappeared in traffic and a cloud of dust and smoke. I sighed and started walking into the building.

Modern Market Guest House

(Quarantine Centre), 15th April 2021

16:00 Hrs.

I went to the third floor. My arms were in pain due to carrying heavy luggage bags. As soon as I got there, the receptionist opened the register and entered my details.

"Room No. 419, sir!" the receptionist handed over to me the room keys.

"Thank you!" I smiled and left for my room.

I was feeling like a newcomer in a boys' hostel. As soon as I opened the door of my room, a hot breeze kissed my face. The room was hot as an oven, and I felt like a chicken inside it.

"Be ready to roast, dear," I said in my mind. The facility was for two persons, but only one person per room was allowed this time as per the new rule. Two beds, one study table, two chairs, new cushions, a new mattress, new bed sheets, two cupboards, two fans that's all I had in my room. I called Nency to ensure that she reached home safely. She insists on me to show my room. After the video call, I checked for drinking water the jugs were empty. I took the jugs, came into the gallery, and asked one of the staff members for the source of drinking water. He showed me a cold drinking water tank located on the third floor. The source of drinking water for all was the same. I thought for a while that what if a COVID-positive person used this water source? I refilled my water jugs and returned to the room.

Due to a change in place, I could not sleep at night. So late at night, about 01:30 a.m., I was sitting in the gallery. It was excellent in the night, and I felt some relief. The party songs were being played on the second floor, and I thought someone was partying late at night. As all young engineers stayed in the guest house, late-night parties were a regular thing. Also, it was too dull and head banger to stay in a room for months and months without doing anything.

I sat for one and a half hours there. Then, finally, late after 03:30 a.m., I tried to sleep, and I succeeded.

16th April 2021, Day -2 At Quarantine.

The morning was not too good. As soon as I woke up, the heart-piercing sirens of ambulances were all over in the atmosphere. The fear was in the air as the virus was airborne. The service boy knocked on the door; he brought snacks and tea. I thanked him. I sipped tea. The tea looked good in color and texture but was terrible in taste. The taste was like sugar syrup.

The lunch was not up to expectations but luckily was eatable. The laundry facility was good this time, but I preferred to wash my clothes. I continued writing my book. The book was at a stage where I was stuck; I could not find a story. I was stressed. I have not smoked even a single time in my life, but I was feeling stressed, so I was looking to smoke, but thanks to the quarantine center, I could not order cigarettes. So, ultimately I survived being a smoker.

My phone rang around 03:00 p.m. I picked the call; the call was from my colleague and a good friend Mr. Kalpesh Dhokia. Kalpesh Ji is in the Mechanical department. Very enthusiastic, technically very sound, and workaholic person. I enjoy his company working with him.

"Hey! Where are you?" Kalpesh Ji asked.

"I am here in the quarantine center," I replied.

"Which quarantine center?" He asked.

"Modern Market Guesthouse!" I replied.

"I am also at the same place, which room no. Dear," he said.

"I am at room no. 419, and you?" I said.

"I am in room no. 206!" he replied.

"Ok! So we will meet in the evening, at tea time!" I said.

I was happy that I had a friend to cut this quarantine period quickly. So I can speak my heart out.

Around 5 p.m., Kalpesh Ji came to my room. I offered him the chair. At the same time service boy brought in the tea.

"Perfect timing!" I said that boy was taking teacups from his hand. We had a long discussion on politics to the origin of Covid-19. The time passed quickly. It was 06:15 p.m. Kalpesh Ji insisted on a round of gallery of the building. The gallery was just attached to the kitchen. People were doing different activities there like rope skipping, walking, gossiping, etc. I also like that young people were utilizing their time in productive activities.

The gallery was full of life—cold breezy wind and the city's main road view. Kalpesh Ji and I talked about the lucky people roaming on the road. People on the road were a bit fortunate to refresh themselves by going to market for a while. There is a bakery across the street. People were buying delicious foodstuff from there. I was feeling jealous and hungry too. The traffic increased on the road as night curfew cut-off time was nearing. There was a night curfew implemented after 7 p.m. The sunlight was decreasing artificial lighting has started growing.

Me and Kalpesh Ji were still in conversation. Kalpesh Ji was worried about his aged parents. Meanwhile, A white, old model ambulance with a blue blinking light on the roof rushed on the main road' When that ambulance passed near from our gallery, my heart sank. My throat dried. I was feeling difficulty breathing for a while. Then, I saw Kalpesh Ji. He was also looking stunned.

It was a hearse, not an ambulance. That hearse was full of human bodies packed in plastic body bags! The dead bodies were of patients who succumbed to the Coronavirus. We were able to peep into the hearse as there were no curtains or stickers on the glass windows of the hearse. However, the windows were big enough to see inside. The hearse was carrying at least five bodies in one turn. We were speechless!

"When will this end?" I asked Kalpesh Ji.

"God knows! These Chinese are responsible for all these blunders!" He sighed.

"Probably the bats which were traded at Wuhan seafood market. But I believe humans are only responsible for this pandemic. There is a variety of vegetarian food available, but these Chinese want to eat weird species. I do not understand what is wrong with Chinese people?" I said with frustration.

"Their funda is simple, eat every living species." Kalpesh Ji told me to check his mobile.

His point of view made the environment lighter.

I replied with a smile!

It has become routine for us to spend two hours in the evening on the balcony.

One day I noticed that a poor boy and his little sister were standing near the bakery. They were selling book stickers to the pedestrians. That poor boy was looking to make some living by selling those stickers. He and his sister had worn the mask not only to protect himself from the Coronavirus but also to save himself from the police. The boy's dilemma was to stay inside the house or go out and earn bread! And such dilemma was not only of him, but the difficulty was of millions of daily wage earners!

A young girl came out of that bakery. She looked pretty in a red t-shirt and blue jeans. Her sports shoes tell the story of the financial status of the family. She offered some snacks and a water bottle to that pair of a poor boy and his sister. He accepted the food and water immediately and gave them to her sister. The poor boy and his sister became happy. That pretty girl started walking towards the car. That poor boy ran to her and offered stickers to the girl in return for her help! That rich girl denied it, but the boy insisted on her. The girl chose two stickers and kept them with her. The brother and sister pair kept watching that girl until she disappeared on the road in the Blue shining BMW seven series car! I was feeling happy. Me and Kalpesh Ji saw each other with a smile.

I called my wife, Nency. I asked her about the wellness of her and our children. My younger son told me to come home; tears rolled over my chicks! My elder boy asked me about my return. I was speechless. To empathize with him, I told him that I would return soon. I told them to be very careful, safe at home.

I could not sleep at night due to the silence and heart-piercing noise of sirens. I was deeply worried inside my heart about my family. Once I had a thought to leave the quarantine center and go back home. I was staying months and months away from the family for job purposes. This experience was teaching me the importance of the family!

17th April 2021

02:25 a.m.

"Mr. Nikunj this way!" A ward boy guided me to the coronavirus patients' ward!

"Mrs. Nency ..." I asked the nurse in complete P.P.E. kit.

She pointed me towards the ward before I completed my sentence.

I rushed to the room! But, unfortunately, the Doctors did not allow me to get inside. I was constrained! I was watching Nency slowly succumbing to the deadly infection of Coronavirus. I was watching helplessly from outside. Her struggle to survive ends in a few minutes.

I screamed her name in pain...!

This scary dream vanished my sleep that night. My throat dried. I was shivering with sweat flowing from head to chest! I was feeling difficulty in breathing! I looked for water, but the jugs were empty. So I took them, opened the door, and breathed heavily! To my surprise, as soon as I came out of the room, I heard loud party music coming from the second floor. So I went to the third floor to refill my water jugs. After filling my jugs, I saw the kitchen door was opened when I was returning. Usually, the door is kept locked, but today, it was open. So I went n to the kitchen. No one was there except a few hungry rats! I was cursing myself for eating such food from such a kitchen. I opened the door and went into the gallery. If I had been a smoker, this would have been a golden time for me to have one cigarette!

The surroundings were silent, with a feeling of pain in the air. The cold breeze was kissing my chicks. Suddenly the same hearse passed by with five to six dead bodies packed in white polythene body bags. I sighed and thanked God for saving me from the deadly coronavirus infection in September 2020.

I just sat there and thought about the people and their families who succumbed to this pandemic, probably a pandemic generated by a human-made or modified virus.

Now, I was highly eager to know about the origin of this virus, and I decided that I would write a book on it sooner or later. But then, I remembered that mystery man who messaged me on my telegram channel, including screenshots and photos of some documents related to the Coronavirus outbreak and had not taken them seriously.

I opened the message in my telegram account and started studying that material. More I started researching the material, and I started exploring more and more facts and studies about this outbreak! My curiosity led me to some groundbreaking truths during my journey of writing this book!

II

The Curiosity

17th April 2021

03:05 a.m.

I opened my phone and typed a message in a social media application, "Who are you? Why did you send me these documents?" As usual no answer from the other side. I sent hundreds of messages to that anonymous person, but no response. I tried to call many times on that number, but the number was no more in existence. The documents had some names I saw for the first time in my life, like Peter Daszak and Shi Jheng Li. The information looked classified and pointed towards one place, Wuhan Institute of Virology!

That night I slept at 04:45 a.m. Several questions were floating in my mind, 'who is Peter Daszak?' 'Who is Batwoman?', 'Why my mind is not ready to believe the zoonosis theory of Covid pandemic?' 'What about the samples collected from the Wuhan seafood market?' 'What is the result of those collected samples?' 'How many and what kind of animals were found infected with this virus? I fell asleep. But before I fell asleep, I made a full-page list of questions whose answers must be found out. I was writing a book in the self-help genre. Once I thought about writing a book related to brutal reality. I was worried if I would fail to find a rhythm to writing such a book on sensitive and full of conspiracy subjects.

I woke up at 8 a.m. the morning went to the window. Beside the quarantine center, there was a pigeon-holed property. I could not see the sun in the morning, but the pigeons gave me full company. My right leg was paining, but I neglected that pain for months. Before breakfast, I took some almonds.

At exactly 08:00 a.m., my door knocked. I opened the door the breakfast delivered and made my face seeing the breakfast. The noise of drilling was annoying me. One of the CCTV cameras was precisely in front of my room. The documents which I was studying last night I reopened them and started reading with having tea. The tea was delicious. I wished I would have taken more than a cup. There was a group photo in which Peter Daszak, Shi Jheng li (Batwoman) Could have been easily recognized in that document. There were some messages written in the Chinese Traditional language. At that, two theories were the famous one that was straightforward and had no conspiracy around: Zoonosis. The other one was messy and full of conspiracy: Lab Leak Theory! It could have been a zoonosis, a direct virus jump from Horse Shoe bats to humans, but it did not satisfy my curiosity.

On the other hand, the Lab leak theory needed a superfine and deep investigation. Such a level of inquiry could not be possible inside China without the support of the Chinese administration and the communist government. U.S.A. demanded such investigation from W.H.O. Donald Trump, the president of America, was very confident about the Lab Leak possibility. In one T.V. interview with journalist Sharri Markson. He said that "He could not reveal beyond a limit as he is not the president of America now." But he was very sure about the data the C.I.A. and other intelligence agencies have collected.

Meanwhile, my friend Kalpesh Ji came to my room. He insisted I have some snacks in his room; I locked my room and went to his room. So we had some snacks and green tea.

"Any news?" he asked me.

"No, nothing! Do you have any news?" I asked him.

"No, same here, nothing to say!" he replied.

"So, we are meeting in the evening at the gallery. I have to continue my writing." I said and left his room as early as possible because while quarantined, we were not allowed to roam in others' rooms.

"Sure, see you at the balcony!" Kalpesh Ji replied.

I got bad news on the same day. My father-in-law was found covid positive two days earlier. But to avoid panic and unwanted traveling in such peak time. His condition was critical as the Oxygen level in his blood was dropping faster. My brother-in-law told me that he is in I.C.U., and his frequently fluctuating oxygen level. It was awful for someone who already has diabetes. I was perturbed for him and my brother. My mind has stopped working, and simultaneously my hands too.

I recommended some ayurvedic medicines for my father-in-law and allopathic medication for faster recovery. Nency called me and asked me to see my father-in-law, but I told her it is not a good idea to travel with such high risk and that too along with two children; She too agreed. I told her I could understand her emotions, but this was not the right time to travel. My inlaws also told her to stay at home and contact her. They will inform us of every detail of his case.

23rd April 2021,

The test day

"Sir, please fill up this form and get ready to give a sample for corona testing." A service boy came to my room and handed me a blank personal details form.

I took the form and fill it up. The testing was completed. We were eager to know the results as there were no restrictions despite being a quarantine site. One more day passed. Me and Kalpesh Ji gathered in the evening at the balcony.

The sky was full of shades of different colors, from Orange to sea blue! The atmosphere was evident due to lockdown. There was less traffic on the roads, so the atmosphere was getting into the actual shape pure and natural due to significantly less pollution. From Punjab, India, the Himalayas have seen many news channels cover that news under the headline, 'Nature is curing itself.' And it was true.

We, humans, are social animals. We love gatherings, partying, socializing. But this pandemic changed everything. We were trapped in our houses, and nature became free. It looked like nature got a new life in this pandemic.

Meanwhile, my eyes were following the black dog on the footpath. The dog was probably in search of food. The pups were following the dog. The puppies were adorable in look. They were like God had splashed black color to those dogs. Those pups remind me of a super hit movie of Hollywood in the 90s. The name of that movie was '101 Dalmashions'. Finally, the dog found something to eat from the dustbin; as soon as the dog started eating the food, the pups began feeding on her! I have been reminded of my childhood. We also had a dog at that time. I love to play with 'Sherni.' I kept her name after the name of the female lioness in the Hindi language. 'Sherni' was a brilliant dog; loyal and adorable! I still remember when 'Sherni' left this world, I was in 11th standard. I ate nothing for one and a half days. She left two adorable pups behind her, 'Denny' & Dolly' they also left when they grew up!

Present Day

Me and Kalpesh Ji were busy in our chat. The same hearse was taking multiple bodies at one time in one turn! It was the second time such a horrible thing I was watching in my life. It was the first time it was an accident that I saw live and just survived. One more day ended. We have become hopeless now. The next day the result of our covid test would be declared. We were scared inside if someone tests positive for covid, there will be chances of others too to test positive in the following test. The reason was common drinking water point and many others!

24th April 2021

The results of the covid test were announced. I went to the receptionist and asked about the results. Two cases were found positive!

"Sir, please go to the room!" the receptionist requested me. I followed his instruction. After the slowdown f the chaos created, I went to the receptionist and asked about the positive person's name.

"Sir, room no 206 and 418 employees found positive!" He replied.

"What? Room no. 206?!" I wanted to confirm because room no. 206 was my friend Kalpesh Ji's room number!

I called Kalpesh Ji, but his number was engaged. I kept trying for 10 minutes; then, I thought he must inform his family about his situation. After half an hour, I got a call from him, and he told me that he was okay right now, and management has told him to live quarantine facility as soon as possible. I asked him about the treatment follow-up by the administration. The company management had refused to give him treatment in the company facility or hospitals where the company had a tie-up! This was the natural face of corporate dictatorship because he found positive inside the company quarantine facility, so it was management's responsibility of his treatment. It was heartbreaking for me. Kalpesh Ji and I wanted to leave the quarantine house, but not like this! Management did not even think about his family even once as they told Kalpesh Ji to go home! What if his family gets infected?

I was also worried about myself too. He and I were in contact all the time. We took snacks together and lots of chats with each other. I also started suspecting myself. I had one safety as I had already been infected by corona in September 2020. So I was assuming the antibodies generated at that time would give a good fight. I did not tell anybody about this; My worry for my father-in-law and brother. On that day, kalpesh Ji left the quarantine facility. On that evening, we came to know that the manager of the quarantine facility found positive three days ago! And still, he was staying in the quarantine facility in a separate room. Kalpesh Ji later confirmed that he met that manager regarding some room issues.

So the days of quarantine extended further for four more days. After two more days, the remaining employees undergo the corona test again. The corona testing is done again; six more employees found covid challenged! Finally, we shifted to the company township in a hurry!

12/09/2021, 19:25 hrs

Sunday

While writing this book, I got a severe injury in my right hand. A deep cut by the edge of a metal sheet at the workplace poured my hand in the blood. I did the first aid; believe me, it was not the case of first aid. If I had gone to the doctor, he could have taken stitches! After first aid, I pulled out my note and pen and started writing. I was happy that I could still write at the same speed.

Anyways! Pain is part of life. My pain was nothing in front of those whose loved ones succumbed to this deadly pandemic. Even I lost a few close relatives too during this pandemic.

My curiosity and interest increased towards this corona pandemic as we have discussed earlier that there were two main theories of the origin of the pandemic: one zoonosis and another lab leak.

The questions must be answered, and many scientists and virologists were seeking the answers. I have started following well-known intellectuals and journalists who were sure about the lab leak theory. Worldwide, a group of people began investigating this conspiracy theory deeply. The result of their efforts floated many proofs those were enough to point out at Wuhan Institute of Virology to held responsible for the brutal pandemic of the century!

"Curiosity is the fuel for Discovery, Inquiry, and Learning!"

III

The Dark History

"The more you know about the past,

The better prepared you are for the future"
-Theodore Roosevelt

21 February 2003

An older adult about 64 years of age in a black coat, Gray pent, and leather handbag was heading towards the Metropole hotel in Hong Kong, South China. The doctor checks in to the 9th floor of the hotel. That older man was coughing casually. He was in Hong Kong to attend a wedding. But, unfortunately, he was going to be a super spreader of a deadly virus that would kill hundreds of people in the upcoming days.

The older man was Professor Liu Jianlun. He worked as a consulting doctor at Sun Yat-sen memorial hospital, Guangzhou. Five patients died due to a mysterious illness in November 2002; from those five patients, the first patient was a farmer from Foshan. Professor Liu was treating the patients of SARS coronavirus patients. The hospital where Liu was working; revealed that the antibodies for the SARS virus do not appear in a healthy human body. The antibodies only occur when the body is fighting with the SARS coronavirus.

23 February 2003,

Hong Kong

Dr. Liu confessed that he was secretly trying to contain a deadly outbreak in southern China. Slowly his lungs were filled with water. In less than a month, professor Liu died. The virus was eating through the patient's lungs. The experts realized that it was not an ordinary virus. Dr. Liu also knew that hundreds of patients treated the mysterious [SARS –CoV] virus! *This information was China's state secret!*

Further investigation on the internet I found more interesting facts.

November 2002

Dr. Klaus Stohr

A mysterious viral disease case started reporting. The first case was reported on the date 16 November 2002.

Dr. Klaus Stohr, a German virologist, and epidemiologist traveled from Geneva to Guangdong on 23 November 2002. Dr. Klaus Stohr was with W.H.O for fifteen years. He was head of the global influenza program and SARS coordinator. The reason for his traveling to Guangdong was to attend a conference. Guangdong was the central area of interest for Dr. Klause as Guangdong was believed as the epicenter of viral diseases. A health official from Guangdong informed Dr. Klaus Stohr that a highly infectious disease is spreading in cities. But Dr. Klaus did not take his information seriously. That anonymous person also told Dr. Klaus that patients were dying and health workers were getting sick. Who was that whistleblower? Whose whistle did not take seriously!

The actual duty of informing W.H.O about this viral disease was Dr. Yi Guan. Dr. Yi Guan was running a W.H.O partner laboratory in Hong Kong. It was such sensitive information, but Dr. Klaus did not share it with Dr. Yi Guan. Dr. Klaus Stohr did not inform Dr. Alan Schnur. At the SARS outbreak, Alan Schnur was Team Leader, Communicable Disease Control, in the WHO Beijing office. Before his 1994 assignment to China, he worked with infectious disease eradication and

Control programs in Africa and Asia, including 12 years in the WHO regional offices in Manila and New Delhi. He has authored or co-authored 13 book chapters and journal articles on smallpox eradication, immunizationPrograms, polio eradication, and Severe Acute Respiratory Syndrome (SARS).

Even nobody tried to re-contact that health official from Guangdong who informed Dr. Klaus Stohr about the spreading infectious disease. It was a shocking and casual approach by a WHO official. WHO has a workforce whose main job is to follow the clues, news, and even rumors about such diseases! According to Guangdong officials, they informed Beijing about this outbreak, but Beijing did not tell WHO.

The outstanding question was **how a virus from bats from a dark cave of Yunnan could travel to humans or animals 1300 kilometers in Guangdong?** That too, without causing even a single case in a village just one kilometer from that cave in Yunnan!

14 May 2021

02:00 a.m. (Approximately)

At late-night around 2:00, I studied bat's behavior and their life. I wanted to find out that 'Are bats visitor mammals like visitor birds?' I did not find anything that can support the Zoonosis theory. I often put my smartphone on focus mode, which is an outstanding feature; one cannot disturb social media and other apps by activating focus mode. Calls only are available facility on focus mode. When the smartphone goes into the focus mode, it feels like I am in the era of the '80s and '90s; only landline and cordless phones were available those too, very rarely.

My phone rang. I picked up my phone and checked my wristwatch; it was 2:20 a.m.! I was surprised the call was from one of my friends.

"Where were you? What were you doing?" he asked me straight forward.

"I was working on my book, and my phone was on focus mode," I replied.

"I texted you multiple times but did not get any answer, so I called you." He said.

"Call and text at this time? Anything serious, are you alright?" I asked him.

"Everything is fine. I called you to tell you something about your book on COVID-19." He said to me.

Only two of my friends were aware bout my book on COVID-19. This is because I had shared my idea with them only.

"Okay, what is its data or any insider information?" I asked him.

"Not a piece of insider information, but it would be beneficial in your book. I have sent you an e-mail with that document and link too. Check them and reply to me," he answered.

"Okay, I will see that and by the way, call you tomorrow. Good night! And thank you for the help." I said to him.

"In friendship, no formalities please, no need to thank me. Good night and see you soon," he replied.

"Good night!" I ended the call.

It was nearly 03:00 a.m. I went to my kitchen to have some fruit juice. We stayed as a bachelor, so we had minimal stuff in the kitchen. The kitchen was empty, but we ensured that the refrigerator was full of fruits, milk, and fruit juices, one more reason to keep the fridge packed with the worst food quality inside township restaurants. Of course, the food was free for us but rarely eatable too. I had some juice and started checking my e-mail box. There was an e-mail with the attachment; I opened the link and attachment. The information was beneficial and relevant to the COVID-19 pandemic too.

According to that report, after the SARS pandemic (2002-2004), a team led by Shi Zhengli and Cui Jie of Wuhan Institute of Virology in China sampled thousands of Horseshoe [Rhinolophus – Biological name] bats!

They found a particular cave in Yunnan, south-western China, where the strains of CORONA virus looked similar to the human version of SARS-CoV. So the researchers spent five years monitoring the bats that lived there, collecting the fresh guano and anal swabs!

They sequenced the genomes of 15 viral strains from the bats and found the make-up of the human version. However, there were no shreds of evidence that the human versions of the SARS strain could jump from bat to another animal.

Cui and Shi searched for other bat populations that could have produced strains capable of infecting humans. As a result, the researchers have isolated more than 300

CORONA virus bat genome sequences, most yet not published! Surprisingly! Also, Shi and her team warned that a deadly outbreak could emerge again!

This sequence of events gives us the reason to look at the Wuhan Institute of Virology and hold responsible for the COVID-19 pandemic!

April 2012, Tongguan, Mojiang, Yunnan Province

Six miners were approaching a mine with shovels in their hands. All of them were going to face the worst phase of their life. The miners were going to open a new chapter in the history of viral diseases!

In that abandoned mine in Tongguan, six mineshaft workers working there were unaware of the population of bats. They were cleaning the way. During that cleaning, there was guano of bats in that waste. The stinky guano of that waste carried the CORONA virus, which was later known as RaTG13! Two weeks (14 days, the period after which the symptoms of CORONA even covid-19 appears) after that mining job, all six workers fell sick with a flu-like disease. These Chinese medical records were kept secret for years. The symptoms were identical to the symptoms of COVID-19. Three of six miners died. One was lingering in the hospital for five months before succumbing to the strange illness!

Dr. Jonathan Latham, Executive Director BioScience project, New York, said it was enough time for Coronavirus to mutate into a variant being passed on to the other humans. Further, he added that "The duration of time that the miners were in the hospital, they were incubating the virus for a very long time. That allowed the virus to evolve and adapt the human lungs." The blood samples of dying miners were sent to the highest security bio lab. One thousand five hundred kilometers to the north, yes! To

Wuhan institute of virology, those tested positive for the Coronavirus. Remember, during the SARS breakout in 2002-2004, Dr. Shi Zhengli and her team collected the bat samples and isolated the genome sequences simultaneously, WIV. The samples collected from the mine were also sent to Wuhan Lab. The comparison test revealed that it was a 96% match.

Dr. Robert Gary is a professor of microbiology at Tulane Medical School, New Orleans. He agrees that WIV was researching bat coronavirus called RaTG13 and samples of dead miners. However, he said that "it is scientifically impossible to have mutated into COVID-19 in the lab. Or in the human body of dying miners! It would take about 50 years of natural evolution! It was a shocking statement and a dead-end to the possible origin of the COVID investigation. Still, the question, "How SARS-CoV was rapidly contagious in humans?" the possible answer can only be found in the WIV database. In the 2002-2004 SARS-CoV pandemic, around 300 bat coronavirus sequences were UNPUBLISHED!

The type of research at WIV raises eyebrows and forces us to think again that 'Is COVID-19 pandemic a Zoonosis?' and adds weight to the laboratory leak possibility.

But before moving ahead,

My curious mind was overflowing with questions. The questions must be answered. Without finding answers to the questions listed below, one cannot understand what happened and the co-relation of history with covid-19!

Question: What were those workers doing there? Why do they go to the particular mine in Mojiang? And most important one for whom they were working?

In 2012-13, a project for research on pathogens and possible pandemics by them was launched. The removal of debris from the abandoned mine in Mojiang to clear the way but in contrast in the virologist community that the actual job of those mine-workers was to collect the guano of bats! The screenshot below shows that the sampling of bats was ongoing in 2012. So, it gives weight to the statement that the miners were working for the scientists working on the bat coronavirus.

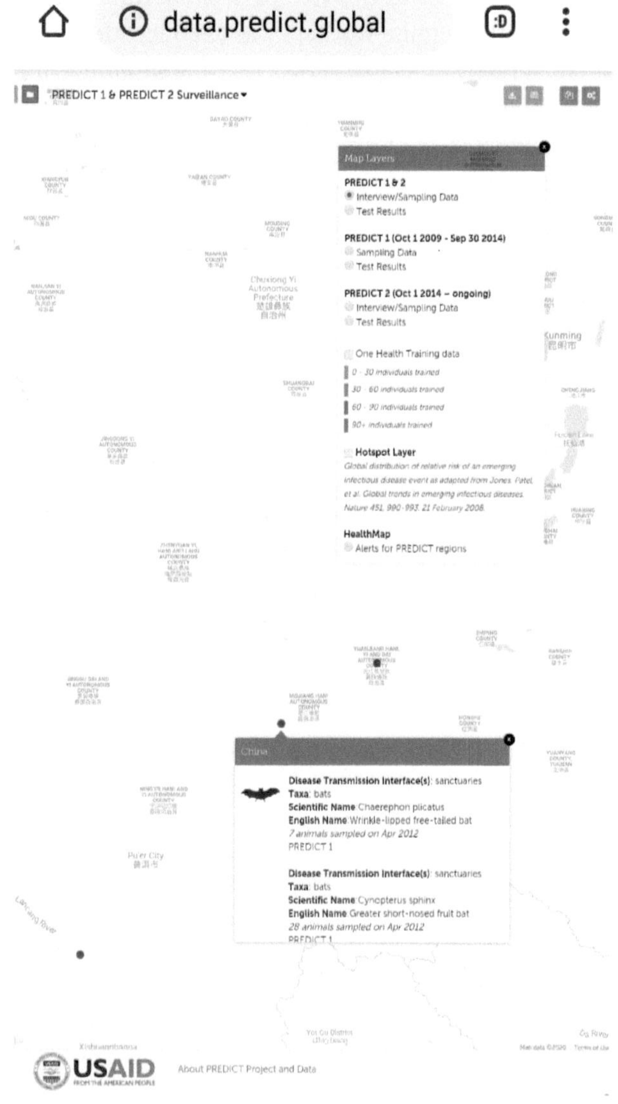

Sampling Data

According to Shi Zhengli, the closest relative RaTG13 of covid-19 was found from the same abandoned mine of Mojiang! The similarity between the RaTG13 virus and SARS-CoV-2 is more than 96% [98.7%-98.9%].

Ultimately solving the mystery of RaTG13 could lead us to conclude what happened and how the COVID-19 pandemic began. However, only considering RaTG13 would not be enough to reach the close of a laboratory leak. Still, as we move further, we explore the possibilities of Laboratory leaks at WIV.

IV

The Epicenter

"I can't change the direction of the wind, but I can adjust my sails always to reach my destination."
-Jimmy Dean

October 2021

I was midway in writing this book. One day I decided to go away from home to write peacefully and non-distracted way. But I was not sure in my mind about the place. So I took my laptop bag, started my scooter, and left. Still have no clue where to go. I kept riding in the city after two crossroads. Finally, I realized that I had reached the place where the seed of this book was planted. The site was that bakery, which Kalpesh Ji and I used to stare at during our quarantine days [Chapter: 1]. I parked my scooter and entered the bakery. I ordered my favorite fast-food items. Open up the laptop; my laptop is a bit slow. Let's come to the point.

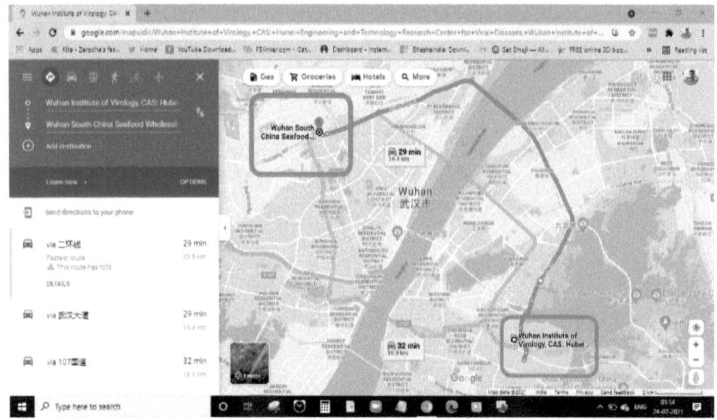

Map of Wuhan (Image Credits Google Map) Figure. 4.1

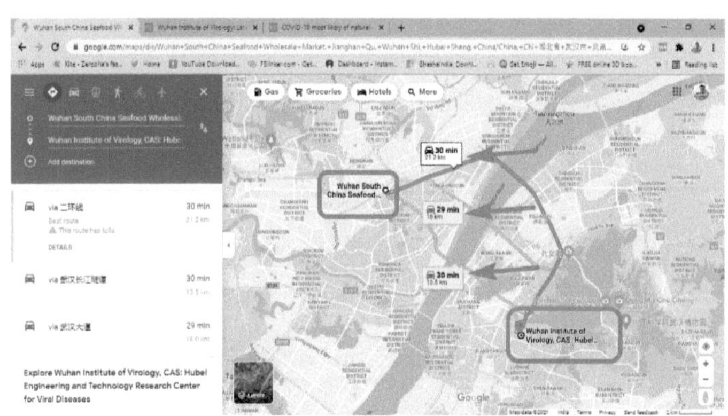

Map Credits: Google maps Figure 4.2

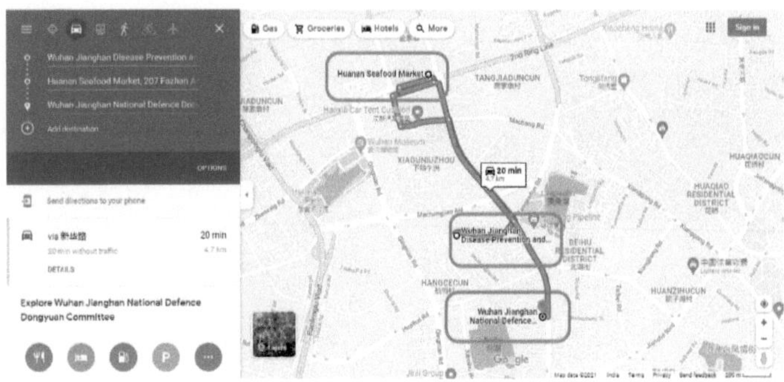

Wuhan Map - 2 (Credits Google Maps) Figure 4.3

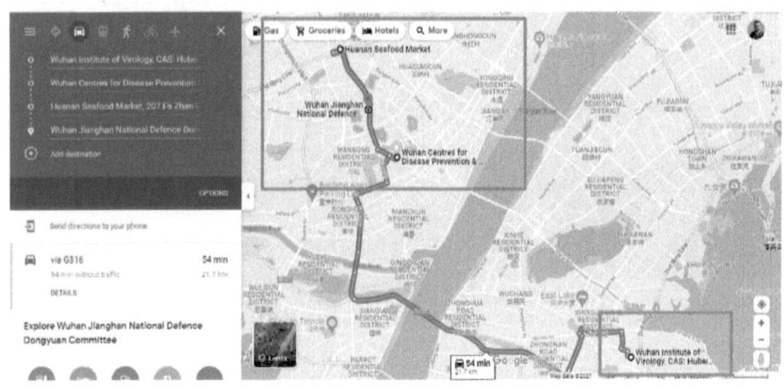

Wuhan Map (Map Credits: Google Maps) Figure 4.4

Please keep in mind above maps throughout the chapter would give you a better understanding of the origin.

SARS –CoV [2002-2004] Zoonosis Origin and Supporting Facts

"I haven't seen anything that makes me feel, as a researcher who studies zoonotic disease, that this market is a likely option," said Colin Carlson, a professor at Georgetown University. Carlson does not work for the WIV.

The theory is plausible that the virus may have jumped from animals to humans. The host animal needs to contact humans, and the virus transmits from that animal to humans. As seen in the chapter 'The Dark History,' the SARS virus in 2002 jumped from bats to the Himalayan Palm civets, Raccoon dogs, then humans. Among them are Palm civets especially. Mostly the SARS isolated from marketplace Palm Civets. The sporadic infections observed in 2003-2004 were associated with the restaurants where palm civet meat was prepared and served. SARS-CoV can persist in palm civets for weeks.

Whereas marketplace animals were the potential immediate source of infection, the palm civets found on a farm and wild were free of infection. Pieces of evidence suggest that civet-like animals can play the role of conduit (vector) for the virus from another precursor or reservoir host. For example, although SARS-CoV antisera and the virus itself were overwhelmingly present in marketplace palm civets in Guangdong, the vast majority of civets on farms and in the wild were found free of infection.

SARS-CoV-2 and Wuhan Wet Market

Some early cases of the outbreak were found associated with Wuhan-Huanan Wet Market. The researchers collected environmental samples which landed on the surface of the market. But tissue samples of animals collected revealed that no trace of the virus. For the virus to jump from animals to humans, animals must be carrying it.

Let's look at the map of Wuhan carefully. We can see that Wuhan Institute of Virology, Wuhan-Huanan Seafood market, Wuhan Centres for Disease Prevention & Control, and Wuhan Huanan National Defence Dongyuan Committee are within one hour distance; in the context of this paragraph will understand in another chapter, 'The Invisible Weapon' in detail.

There are many controversies about the first case; officially, the first case was reported in December 2019. And there are also believers that the first case appeared in November 2019, and by January, it spread over globally!

Initially, the focus was on the wild animal trade at the Huanan Seafood Market, where the first reported cases were publicly confirmed. However, the earliest publicly reported case using unclassified data had no exposure to the Huanan Seafood Market. Instead, it developed symptoms on December 1, 2019, nine days before the first patient connected to the market. In addition, the virus strains sampled in the market were shown to be already adapted to human transmission, indicating that the virus had jumped species earlier.

On May 15, the Bulletin of the Atomic Scientists published a short commentary titled, "Let evidence, not talk radio, determine whether the outbreak started in a lab," by Ali Nouri, a biologist and President of the Federation of American Scientists. "The outbreak" referred to the pandemic of SARS-CoV-2 now circling the globe. It is a thin commentary, and it is puzzling why the Bulletin thought it desirable to publish it at all. Two weeks earlier, the journal had published a reasoned and competent appraisal by Kings College London biosecurity expert Filippa *Lentzos* titled, "Natural spillover or research lab leak? Why a credible investigation is needed to determine the origin of the coronavirus pandemic."

The most damaging blows in search of the origin of Covid-19 were the preannouncements from the President of the USA, Donald Trump. But long before Trump, Pompeo, and Co. sought a Chinese scapegoat for the President's gross and willful incompetence, researchers understood that the possibility of laboratory escape of the pathogen was plausible, if unproven, case. It is most definitely not "a conspiracy theory."

As we can see in the map images, Wuhan has two virology institutes to consider, not one: The Wuhan Centre for Disease Control and Prevention (WHCDC) and the Wuhan Institute of Virology (WIV). Both have conducted large projects on novel bat viruses and maintained extensive research collections of novel bat viruses. At least the WIV possessed the virus that is the most closely related known virus globally to the outbreak virus, bat virus RaTG13. This virus was isolated in 2013 and had its genome published on January 23, 2020. Seven more years of bat coronavirus collection followed the 2013 RaTG13 isolation.

Official Chinese government recognition early in the SARS-CoV-2 outbreak of biosafety inadequacies in China's high containment facilities. In February 2020, several weeks after the COVID-19 outbreak in Wuhan, China's President Xi Jinping stressed the need to ensure "biosafety and biosecurity of the country." On January 1, Wuhan Institute of Virology's director-general, Yanyi Wang, messaged her colleagues, saying the National Health Commission told her the Lab's COVID-19 data shall not be published on social media and shall not be disclosed to the press. And on January 3, the commission sent this document, never posted online, but saved by researchers, telling labs to destroy COVID-19 samples or send them to the depository institutions designated by the state. Late Friday [May 16, 2020], the Chinese government admitted to the destruction ... but she said it was for public safety.

Shi Zheng-Li took the possibility of a laboratory escape perfectly seriously. Jonna Mazat of the University of California-Davis, a collaborator with Dr. Shi, told Josh Rogin of the Washington Post, "Absolutely, accidents can happen." In an interview with Scientific American, Shi admitted that her very first thought was, "If coronaviruses were the culprit, she remembers thinking, 'Could they have come from our lab?"

It is important to note that no intermediary host has yet been identified for the SARS-CoV-2 virus. The authors also stated that "No animal sampling prior to the shutdown and sanitization [of the Wuhan fish market] was done." The Chinese government scrapped the previous official story about the Wuhan fish market on May 26. China's top epidemiologist said Tuesday that testing samples from a Wuhan food market, initially suspected as a path for the virus's spread to humans, failed to show links between animals being sold there and the pathogen. Gao Fu, director of the Chinese Center for Disease Control and Prevention, said, in comments carried in China state media. No SARS-CoV-2 isolates were detected in animals or fish sold at the market, only in environmental samples, including sewage. Gao Fu added, "At first, we assumed the seafood market might have the virus, but now the market is more like a victim. The novel coronavirus had existed long before." The same statement was published in South China Morning Post on February 24, 2020.

Map, WHO, and First Case

The detailed analysis of the WHO map and COVID-19 case 'ZERO' was published on zenodo.org with the pen name of the Author Chang, Whatsin Wu. It ultimately becomes Whatsin Wuchang!

World Health Organization's report on SARS-2 contains several maps about the first published COVID-19 patient. The first map in the `Spatial distribution' section of the Annex features two visible dots on the west side of the Yangtze river in Jianghan district. The description above refers to this map as \The first stage of onset: 8-11 December 2019, cases were sporadic". The two dots and the description suggest that the map includes the first published cases, one who fell ill on December 8, 2019, and another who fell sick on December 11, 2019.

The investigative report by Eva Dou and Emily Rauhala of the Washington Post showed that the World Health Organization's map erroneously indicated that the first patient lived in Jianghan District, on the west side of the Yangtze River. But actually, the first patient lived in the Wuchang district, which is on the east side of the river.

Moreover, the World Health Organization's report map has several other deficiencies. The map's resolution is abysmal, and the dots showing the location of patients are hard to recognize. For example, it is impossible to discern the third dot on the map even though the legend implies that one may be present. In addition, the label on the map says "December 11", although the description above refers to "8-11 December 2019" cases, and the map shows two dots.

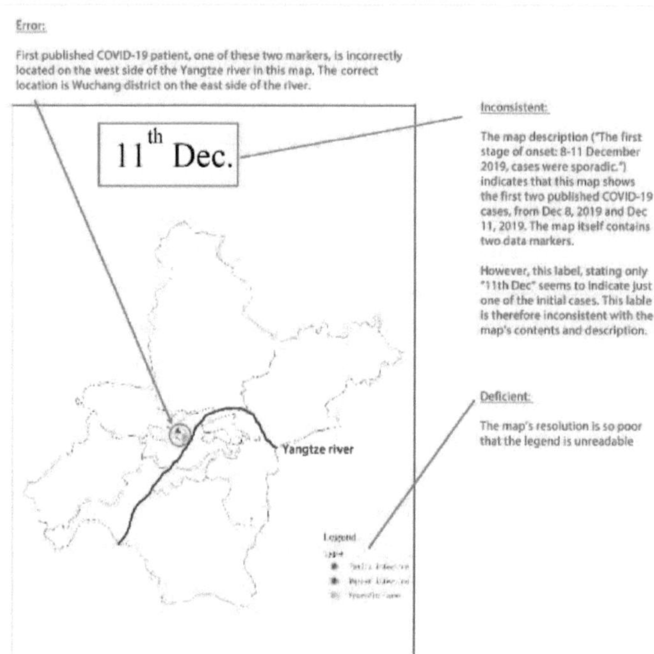

Image Source Credits: https://zenodo.org/record/5202258#.YbbwqNBBzIV

A recent paper by several prominent virologists maps the first two published COVID-19 patients' home addresses, placing the December 8, 2019, and December 11, 2019 patients in Jianghan district on the west side of the Yangtze River. However, this is incorrect because the December 8, 2019 case's home address was in Wuchang district, on the east side of the river.

The Author's error in the map also excludes the location of Wuhan Institute of Virology's headquarters, which is in the same district Wuchang, the district of the first COVID-

19's patient, how such a critical location was not mentioned in the report!

The WHO did not explain why a map in the annexes of the WHO-China joint report appears to show the first case on one side of the Yangtze River, while the Wuhan government had announced last year that the first patient, who fell ill December 8, 2019, lived on the other side of the river, in Wuchang district.

Suppression of information and individuals by Chinese authorities. Two Chinese University academics published discussed both the WHCDC and the WIV and concluded that "the killer coronavirus probably originated from a laboratory in Wuhan"; the publication was removed from the internet by Chinese government officials. The paper had been posted on Research Gate but was blocked after 24 hours. After being placed on an archive file by internet users, it was again blocked after a week, and the two Chinese authors were pressured to retract the paper. However, it is still available on Web archives. Luckily I got the screenshots of that document from web archives. Later the PDF document was also obtained. See the below images.

The possible origins of 2019-nCoV coronavirus

Botao Xiao[1,2*] and Lei Xiao[3]

[1] Joint International Research Laboratory of Synthetic Biology and Medicine, School of Biology and Biological Engineering, South China University of Technology, Guangzhou 510006, China

[2] School of Physics, Huazhong University of Science and Technology, Wuhan 430074, China

[3] Tian You Hospital, Wuhan University of Science and Technology, Wuhan 430064, China

* Corresponding author: xiaob@scut.edu.cn

Tel / Fax: 86-20-3938-0631

Figure 4.5

The 2019-nCoV coronavirus has caused an epidemic of 28,060 laboratory-confirmed infections in human including 564 deaths in China by February 6, 2020. Two descriptions of the virus published on Nature this week indicated that the genome sequences from patients were 96% or 89% identical to the Bat CoV ZC45 coronavirus originally found in *Rhinolophus affinis* [1,2]. It was critical to study where the pathogen came from and how it passed onto human.

An article published on The Lancet reported that 41 people in Wuhan were found to have the acute respiratory syndrome and 27 of them had contact with Huanan Seafood Market [3]. The 2019-nCoV was found in 33 out of 585 samples collected in the market after the outbreak. The market was suspicious to be the origin of the epidemic, and was shut down according to the rule of quarantine the source during an epidemic.

The bats carrying CoV ZC45 were originally found in Yunnan or Zhejiang province, both of which were more than 900 kilometers away from the seafood market. Bats were normally found to live in caves and trees. But the seafood market is in a densely-populated district of Wuhan, a metropolitan of ~15 million people. The probability was very low for the bats to fly to the market. According to municipal reports and the testimonies of 31 residents and 28 visitors, the bat was never a food source in the city, and no bat was traded in the market. There was possible natural recombination or intermediate host of the coronavirus, yet little proof has been reported.

Was there any other possible pathway? We screened the area around the seafood market and identified two laboratories conducting research on bat coronavirus. Within ~280 meters from the market, there was the Wuhan Center for Disease Control & Prevention (WHCDC) (Figure 1, from Baidu and Google maps). WHCDC hosted animals in laboratories for research purpose, one of which was specialized in pathogens collection and identification [4-6]. In one of their studies, 155 bats including *Rhinolophus affinis* were captured in Hubei province, and other 450 bats were captured in Zhejiang province [4]. The expert in collection was noted in the Author Contributions (JHT). Moreover, he was broadcasted for collecting viruses on nation-wide newspapers and websites in 2017 and 2019 [7,8]. He described that he was once by attacked by bats and the blood of a bat shot on his skin. He knew the extreme danger of the infection so he quarantined himself for 14 days [7]. In another accident, he quarantined himself again because bats peed on him. He was once thrilled for capturing a bat carrying a live tick [8].

Surgery was performed on the caged animals and the tissue samples were collected for DNA and RNA extraction and sequencing [4,5]. The tissue samples and contaminated trashes were source of pathogens. They were only ~280 meters from the seafood market. The WHCDC was also adjacent to the Union Hospital (Figure 1, bottom) where the first group of doctors were infected during this epidemic. It is plausible that the virus leaked around and some of them contaminated the initial patients in this epidemic, though solid proofs are needed in future study.

The second laboratory was ~12 kilometers from the seafood market and belonged to Wuhan Institute of Virology, Chinese Academy of Sciences [1,9,10]. This laboratory reported that the Chinese horseshoe bats were natural reservoirs for the severe acute respiratory syndrome coronavirus (SARS-CoV) which caused the 2002-3 pandemic [9]. The principle investigator participated in a project which generated a chimeric virus using

Figure 4.6

the SARS-CoV reverse genetics system, and reported the potential for human emergence [10]. A direct speculation was that SARS-CoV or its derivative might leak from the laboratory.

In summary, somebody was entangled with the evolution of 2019-nCoV coronavirus. In addition to origins of natural recombination and intermediate host, the killer coronavirus probably originated from a laboratory in Wuhan. Safety level may need to be reinforced in high risk biohazardous laboratories. Regulations may be taken to relocate these laboratories far away from city center and other densely populated places.

Contributors

BX designed the comment and performed literature search. All authors performed data acquisition and analysis, collected documents, draw the figure, and wrote the papers.

Acknowledgements

This work is supported by the National Natural Science Foundation of China (11772133, 11372116).

Declaration of interests

All authors declare no competing interests.

References

1. Zhou P, Yang X-L, Wang X-G, et al. A pneumonia outbreak associated with a new coronavirus of probable bat origin. *Nature* 2020. https://doi.org/10.1038/s41586-020-2012-7.
2. Wu F, Zhao S, Yu B, et al. A new coronavirus associated with human respiratory disease in China. *Nature* 2020. https://doi.org/10.1038/s41586-020-2008-3.
3. Huang C, Wang Y, Li X, et al. Clinical features of patients infected with 2019 novel coronavirus in Wuhan, China. *The Lancet* 2019. https://doi.org/10.1016/S0140-6736(20)30183-5.
4. Guo WP, Lin XD, Wang W, et al. Phylogeny and origins of hantaviruses harbored by bats, insectivores, and rodents. *PLoS pathogens* 2013; **9**(2): e1003159.
5. Lu M, Tian JH, Yu B, Guo WP, Holmes EC, Zhang YZ. Extensive diversity of rickettsiales bacteria in ticks from Wuhan, China. *Ticks and tick-borne diseases* 2017; **8**(4): 574-80.
6. Shi M, Lin XD, Chen X, et al. The evolutionary history of vertebrate RNA viruses. *Nature* 2018; **556**(7700): 197-202.
7. Tao P. Expert in Wuhan collected ten thousands animals: capture bats in mountain at night. *Changjiang Times* 2017.
8. Li QX, Zhanyao. Playing with elephant dung, fishing for sea bottom mud: the work that will change China's future. *thepaper* 2019.
9. Ge XY, Li JL, Yang XL, et al. Isolation and characterization of a bat SARS-like coronavirus that uses the ACE2 receptor. *Nature* 2013; **503**(7477): 535-8.
10. Menachery VD, Yount BL, Jr., Debbink K, et al. A SARS-like cluster of circulating bat coronaviruses shows potential for human emergence. *Nature medicine* 2015; **21**(12): 1508-13.

Figure 4.7

Figure 1. The Huanan Seafood Market is close to the WHCDC (from Baidu and Google maps).

Figure 4.8

If the above documents are not readable, I am giving a summary here. The above documents state that *the SARS nCoV-2019 was found in 33 out of 585 samples collected in the market after the outbreak. The market was suspicious of the epidemic's origin [now it turned into a pandemic] and was shut down.*

The bats carrying CoV ZC45 were initially found in Yunnan or Zhejiang province, both of which were more than 900 kilometers away from the Wuhan Seafood Market. Bats are typically found to live in caves and trees. But the seafood market is in the densely-populated district of Wuhan, a metropolitan of ~ 15 million people. Therefore, the probability was very low for the bats to fly to the market. According to municipal reports and testimonies of 31 residents and 28 visitors, ***the bat was never a food source in the city, and no bat was traded in the market. There was possible natural recombination or an intermediate host of the coronavirus, yet little proof has been reported.***

Dragon T.V. Global Intersection released an official video on YouTube on December 11, 2019, under Wilderness Youth (Wilderness Youth | The Invisible Line of Defense (English Subtitled Version). In this video, a scientist Tian Junhua, Associate Chief Technician, Department of Disinfection and Pest Control, Wuhan Municipal center for Disease Control and Prevention. His work is virus sample collection and classification. He explains that bats are rich in various viruses inside all known creatures. In this video, we can also recognize the bats are urinating! Moreover, in the above document, there is mentioned that *'In another incident, he quarantined himself again because bats peed on him.'*

The Author of the report further writes,

Was there any other possible pathway? We screened the area around the seafood market and identified that two laboratories were researching the bat coronavirus. Within ~280 meters of the seafood market, there was Wuhan Centre for Disease Control and Prevention. (WHCDC)[Figure 4.8] *and the other laboratory is Wuhan Institute of Virology.* Wuhan Institute of Virology is not only a research institute but is also a bank of viruses! It is an open truth.

The Chinese keyboard warriors deliberately edited and spread over the internet some videos of PASAR EXTREME MARKET Langowan, North Indonesia, showing those videos as Wuhan Seafood market and spread over social media platforms. Upon comparing the falsely edited videos with actual videos of Pasar Extreme Market, Langowan; confirmed that the videos were falsely distributed to boost the Zoonosis theory. So, most of the world's population believed that the virus had emerged from bats sold at Wuhan Seafood Market. See below the map.

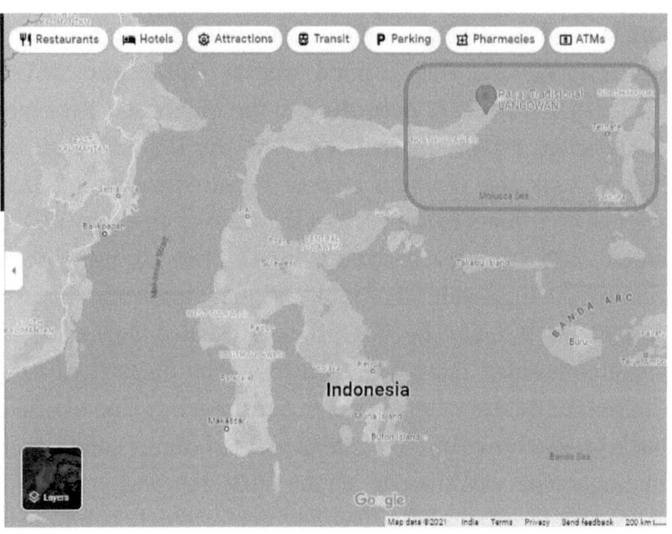

Langowan Indonesia [Credits Google Maps]

Live Bats at Wuhan Institute of Virology

In a media interview, the director of Wuhan Institute of Virology, Yanyi Wang, denied having the live virus in their laboratory before the Covid-19 outbreak. Peter Daszak, under whose leadership and funding of the EcoHealth Alliance, researched coronaviruses in collaboration with Shi ZhengLi at Wuhan Institute of virology denied that Wuhan Institute of Virology was not culturing live bats.

One more thing to be considered here the BSL-4 laboratory was fully operational at Wuhan Institute of Virology in the year 2018. High-risk research on pathogens like Gain-of-Function must be carried out at the BSL-4 laboratory. But, according to an NIH report, the project '*Understanding The Risk of Bat Coronaviruses*' was launched and started in 2014. So, we can conclude that such a high-risk experiment was carried out at BSL-2 and BSL-3 facilities before 2018. According to the highly sensitive but unclassified report, In January 2018, the U.S. Embassy in Beijing took the unusual step of repeatedly sending U.S. science diplomats to the Wuhan Institute of Virology (WIV), which had in 2015 become China's first laboratory to achieve the highest level of international bioresearch safety (known as BSL-4). The U.S. officials learned during their visits concerned them so much that they dispatched two diplomatic cables categorized as Sensitive But Unclassified back to Washington. The cables warned about safety and management weaknesses at the WIV lab and proposed more attention and help. The first cable, which I obtained, also warns that the Lab's work on bat coronaviruses and their potential human transmission represented a risk of a new SARS-like pandemic.

In June 2021, video footage from WIV's BSL-4 facility was published by a group of independent Lab. Leak investigating team DRASTIC. In that 10 minutes video, there was a separate facility to store living bats. The video was in 2017, just before the BSL-4 facility became fully operational in 2018! In that video, the scientists or trained workers were seen feeding bats!

As YanYi Wang denied in an interview with CGTN having live samples of coronaviruses, she probably forgot to say that they had the living source of various coronaviruses!

In an excellent review published in February 2019, Lynn Klotz of the Center for Arms Control and Non-Proliferation noted that three releases of Ebola and Marburg viruses from BSL-4 and lower-containment facilities in the United States had occurred due to incomplete inactivation of cultures. Releases via infection of researchers took place in the highest containment facilities in the United States—at the Centers for Disease Control and Prevention (CDC) in Atlanta and the U.S. Army Medical Research Institute of Infectious Diseases (USAMRIID)—but in all cases, only the researcher became ill. There was no further transmission of the pathogen.

The relevant information is again available from Chinese and Western sources regarding the Wuhan Institute of Virology. Data from official Chinese government sources appeared in a Voice of America report which noted:

There is Chinese evidence that the Lab had safety problems. VOA had located state media reports showing that national inspections flagged security incidents and reported accidents when workers caught bats for study. About a year before the coronavirus outbreak, a security review conducted by a Chinese national team found the Lab did not meet federal standards in five categories. The document on the Lab's official website said after a rigorous and meticulous review, the team gave a high evaluation of the Lab's overall safety management. "At the same time, the review team also put forward further rectification opinions on the five non-conformities and two observations found during the review."

In addition to problems in the Lab, state media also reported that national reviewers found scientists were sloppy when they were handling bats. For example, one of the researchers working at the Wuhan Center for Disease Control & Prevention described to China's state media that bats once attacked him, and he ended up getting bat blood on his skin. *See Figure 4.6.*

In another incident, the same researcher forgot to take protective measures, and the urine of a bat dripped "like rain onto the top of his head," reported China's Xinhua state news agency.

According to the project report of NIH [The National Institute of Health, USA], Project Number 2R01AI110964-06 was launched in the leadership of Peter Daszak. Awarded organization Eco Health Alliance to understand **the Risk of Bat Coronavirus Emergence Novel zoonotic**, bat-origin CoV are a significant threat to global health and food security, as the cause of SARS in China in 2002, the ongoing outbreak of MERS, and a newly emerged Swine Acute Diarrhea Syndrome in China. In a previous R01, we found that bats in southern China harbor an extraordinary diversity of SARSr-CoVs, some of which can use human ACE2 to enter cells, infect humanized mouse models causing SARS-like illness, and evade available therapies or vaccines. Furthermore, we found that people living close to bat habitats are the primary risk groups for spillover. For example, at one site, diverse SARSr-CoVs contain every genetic element of the SARS-CoV genome and identified serological evidence of human exposure among people living nearby. These findings have led to 18 published peer-reviewed papers, including two articles in Nature and a review in Cell. Yet salient questions remain on the origin, diversity, capacity to cause illness, and risk of spillover of these viruses. This R01 renewal will address these issues through 3 specific aims:

Aim 1. Characterize the diversity and distribution of high spillover-risk SARSr-CoVs in bats in southern China. We will use phylogeographic and viral discovery curve analyses to target additional bat sample collection and molecular CoV screening to fill in gaps in our previous sampling and fully characterize natural SARSr-CoV diversity in southern China. In addition, we will sequence receptor binding domains (spike proteins) to identify viruses with the highest potential for spillover, which we will include in our experimental investigations.

Aim 2. Community and clinic-based syndromic surveillance to capture SARSr-CoV spillover, routes of exposure, and potential public health consequences. We will conduct biological-behavioral surveillance in high-risk populations, with known bat contact, in community and clinical settings to 1) identify risk factors for serological and PCR evidence of bat SARSr-CoVs; & 2) assess possible health effects of SARSr-CoVs infection in people. We will analyze bat-CoV serology against human-wildlife contact and exposure data to quantify risk factors and health impacts of SARSr-CoV spillover.

Aim 3. In vitro and in vivo characterization of SARSr-CoV spillover risk, coupled with spatial and Phylogenetic analyses to identify the regions and viruses of public health concern. We will use S protein sequence data, infectious clone technology, in vitro and in vivo infection experiments, and receptor binding analysis to test the hypothesis that % divergence thresholds in S protein sequences predict spillover potential. We will combine these data with bat host distribution, viral diversity and phylogeny, the human survey of risk behaviors and illness, and serology to identify SARSr-CoV spillover risk hotspots across southern China. Together these data and analyses will be critical for the future development of public health interventions and enhanced surveillance to prevent the re-emergence of SARS or the emergence of a novel SARSr-CoV.

In simple language, Aim 3 states that de novo synthesis is used to construct a series of novel chimerical viruses, comprising recombinant hybrids using different spike proteins from each of a series of unpublished natural coronaviruses in an otherwise-constant genome of a bat coronavirus. The ability of the resulting novel viruses to infect human cells in culture and to infect laboratory animals would be tested. The underlying hypothesis is that a direct correlation would be found between the receptor-binding affinity of the spike protein and the ability to infect human cells in culture and to infect laboratory animals. This hypothesis would be tested by asking whether novel viruses encoding spike proteins with the highest receptor-binding affinity have the highest ability to infect human cells in culture and laboratory animals.

However, On November 13, 2019, China's National Health Commission (NHC) issued new influenza guidance instructing hospitals to check patients' blood-oxygen levels with respiratory disease-related pneumonia and look for COVID-19 symptoms. These included: pneumothorax and mediastinal emphysema; acute necrotizing encephalopathy; and multifocal brain damage, including the bilateral thalamus and white matter around the ventricle.

An expert group drafted the influenza plan and medical staff handbook under China's NHC, headed by Wang Chen, with SARS expert Zhong Nanshan as a consultant26. Other members include respiratory disease specialists from various institutes. The social vaccination idea later encouraged by the WHCDC27 was promoted at the group's press conference28. It enables frequent hand washing, avoiding touching the face, and wearing masks. This concept was supported by guidance from the Hubei CDC, which on December 20 issued instructions29 encouraging frequent hand washing, avoiding crowded places, outdoor physical exercise, and advising those with flu-like symptoms to wear masks to prevent transmission to other family members and seek medical attention if symptoms continue to develop.

Soon after this guidance was issued, pneumonic influenza outbreaks and overcrowded hospitals were reported in Hubei and across China. The WHCDC publicly refuted rumors of an influenza outbreak on December 20. This was during a substantial and early spike in officially reported influenza cases.

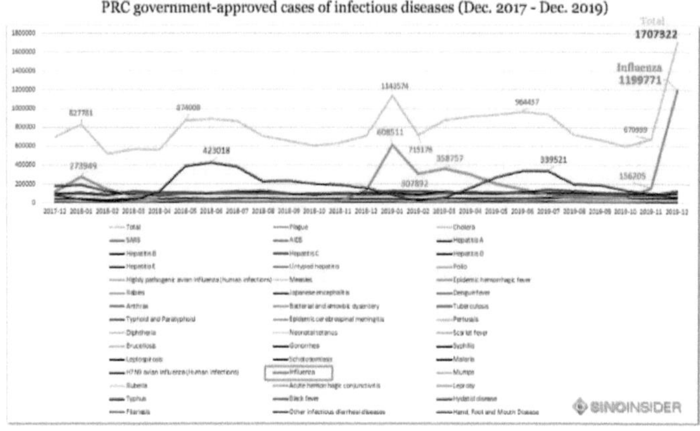

Image Credits [Sino Insider]

An examination of official Chinese data shows a spike in the flu cases in mainland China in late 2019. According to the China Centers for Disease Control and Prevention (China CDC), in December 2019, the number of flu cases had jumped to 1,199,771, nearly twice the 608,511 people who got the flu that January, China's peak for the last winter. Moreover, the month-on-month increase from November 2019 was 768 percent, compared with January's 467-percent rise from December 2018, when there were 130,442 cases recorded.

Like the previous year, October 2019 saw 1.1 times more flu cases than the last month. But November 2019 saw the number of cases triple from October compared with 2018's 1.82-factor increase between October and November. This discrepancy became even more pronounced in December 2019, when flu cases exceeded the November number by 7.68 times.

Flu cases in China typically peak in January. For instance, there were 273,939 flu cases in January 2018 and 120,800 cases in December 2017, that winter's peak. The month-on-month increase was 227 percent. There were 608,511 flu cases in January 2019 and 130,442 cases in December 2018, that winter's peak. The month-on-month increase was 466 percent. By comparison, the month-on-month increase in November 2019 and December 2019 was 768 percent—an atypical phenomenon.

Like the previous year, October 2019 saw 1.1 times more flu cases than the last month. But November 2019 saw the number of cases triple from October compared with 2018's 1.82-factor increase between October and November. This discrepancy became even more pronounced in December 2019, when flu cases exceeded the November number by 7.68 times.

On December 27, 2019, the Influenza Center released its report for the week of Dec. 16-Dec. 22. After this, the weekly flu statistics reports stopped. And beginning in the fourth week of 2020, no more flu outbreaks (10 or more cases in the same community or work unit) were registered.

Based on the timing of the COVID-19 outbreak, we have reason to believe that the uncharacteristic explosion of flu cases in December may be due to the undetected spread of the coronavirus. In reviewing Chinese flu statistics, papers released by Chinese and international health experts, and official PRC statements, we estimate that tens of thousands of people could have been infected with COVID-19 by the end of December.

On January 20, 2020, the PRC acknowledged that the novel coronavirus was contagious among humans after weeks of downplaying the seriousness of the outbreak that started in Wuhan in late 2019. However, COVID-19 had spread to all parts of China by the end of January, and cases were reported in multiple countries.

There are multiple indications that the novel coronavirus outbreak began in November 2019 or even earlier, not in December as initially claimed by the CCP regime.

On February 26, 2020, mainland Chinese media outlet Caixin reported on a paper called "Analysis of mutation and evolution in the coronavirus SARS-CoV-2," released by the Biosafety-Level-3 (P3) laboratory at the Southern Medical University in Guangzhou. According to the paper, the novel coronavirus made its first appearance as early as September 23, 2019.

On March 13, the South China Morning Post reported on an internal PRC government document containing details about the early spread of COVID-19.

The internal document revealed that a 55-year-old Wuhan resident had contracted the disease by November 17, 2019, becoming the earliest known case so far. From that date, there were new cases reported nearly every day. By December 15, 2019, 27 people had been confirmed infected. December 17, 2019, saw the first double-digit increase; by December 20, the number of confirmed cases had reached 60.

The significant number of influenza cases in November 2019 has led some observers to suspect that COVID-19 was already beginning to spread that month. The sharp rise in the stock value of a bio pharmacy company that would work with a PRC government research institute to produce the first COVID-19 vaccine approved for clinical trials is also suspicious. The company, *CanSino Biologics*, saw its stock value more than double from what it was in August.

CanSino Web Page [Cansinotech.com]

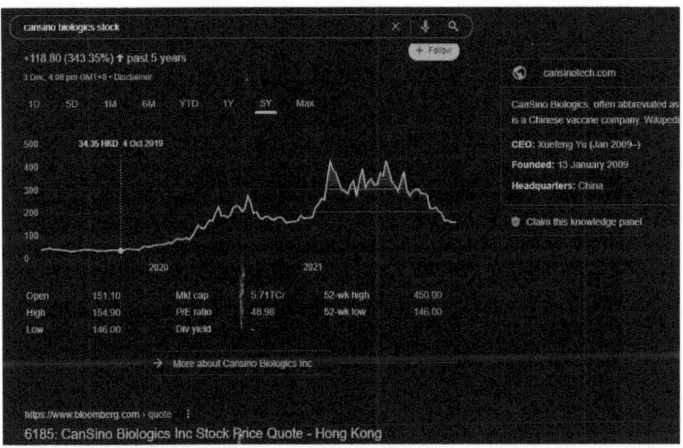

Cansino Tech October-2019 Stock Price

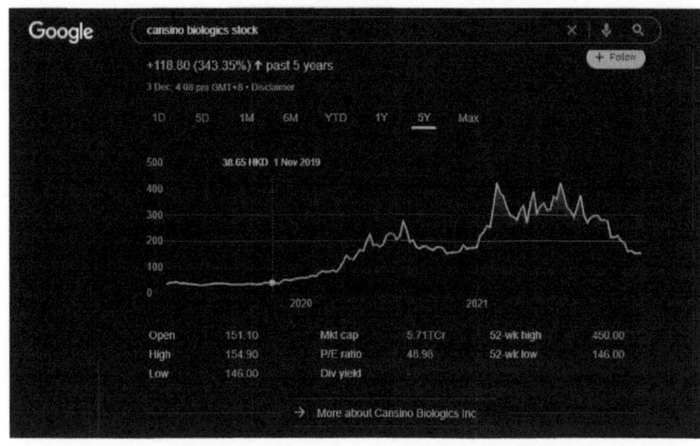

CanSino Stock November-2019 Image Source: Google

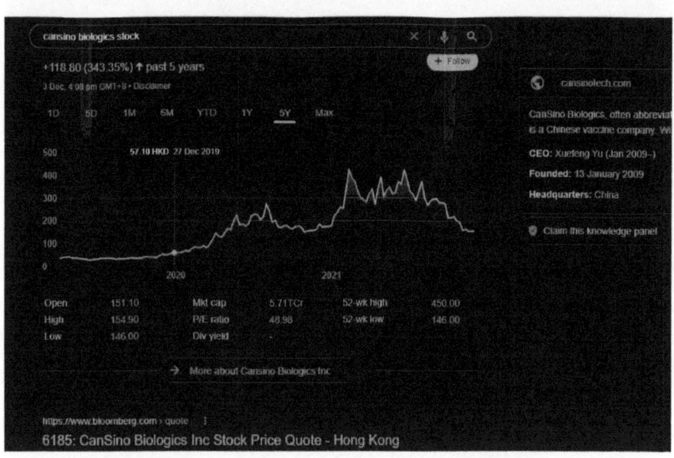

CanSino Tech December-2019 Stock Price

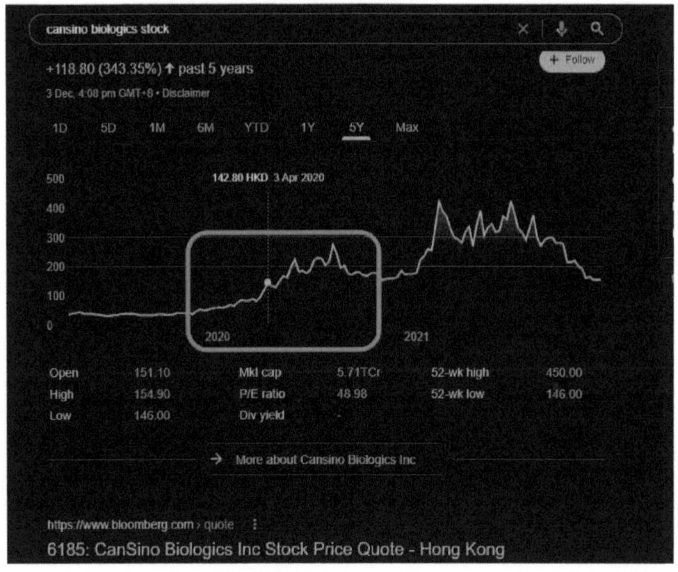

CanSino Stock Sharp Peak From November-2019 to April-2020

The above data confirm that it all started in China much earlier in October- November -2019, but it may be due to the pneumonia-like symptoms; it was not recognized as a COVID-19 outbreak! Also, the graph of *CanSino Biologics* indicates that the vaccine production also would have started much earlier in China, around November-2019.

The graph speaks itself more than words!

Wuhan and Death Toll

On January 28, the Wuhan Municipal Bureau of Civil Affairs announced that it would suspend charges for the cremation of those who died from the new coronavirus (including those suspected of having died because of the virus). The Bureau also requested support for moving corpses (including more funeral vehicles) and additional staffing and protective gear from the Wuhan epidemic prevention command and the Hubei provincial civil affairs department.

Data released by the Hubei Health Commission, as of the end of January 29, Hubei Province had 4,586 confirmed coronavirus cases and 162 deaths; Wuhan City saw 2,261 confirmed cases (49.3 percent of confirmed cases in Hubei) and 129 deaths (79.6 percent of fatalities in Hubei). The provincial health authorities are also tracking 28,780 people who have been in close contact with infected persons, and 26,632 people have been placed under medical observation.

In studying official figures provided by the CCP regime, other open-source information, and information circulated by Chinese citizens on social media, there is a solid reason to believe that the regime has been grossly underreporting the number of confirmed cases and deaths caused by the new coronavirus. By conservative estimates, it is believed that the actual death rate is 35 times or greater than the official figure.

A survey of the publicly available information on virus believed and cremation in Wuhan paints a telling picture:

On January 20, the CCP officially acknowledged the existence of a new coronavirus that can be passed from person-to-person and that Wuhan is ground zero of the epidemic. The Wuhan authorities said that four people had died from the virus; the authorities ordered a lockdown of the city three days later. On January 29, the Wuhan authorities reported 129 virus deaths—a suspiciously low figure in considering other information.

According to reports in overseas Chinese language media, funeral homes in Wuhan operate their cremation furnaces for 24 hours a day; previously, cremations were carried out only afternoon every day. Funeral homes are also reportedly short of funeral vehicles, personnel, and protective gear.

China's overcrowded hospitals and corpses packed in body bags were shown in the videos posted by the citizen journalists like Chen Quishi, Fang, Zhang Zhan, and many more. In one video, Fang says, "This is not a natural disaster but even more so a man-made one. They started covering it up, and they muffled Dr. Li Wenliang, to tell the truth (Govt. authorities). They reprimanded him."

According to other Chinese language media interviews and disclosures on social media, a total of 700 corpses from local hospitals were "disposed of" by funeral homes on January 24; it is unclear how many of the "disposed" corpses belonged to victims of the coronavirus. According to a social media post published by a person claiming to be a worker at the Qingshan Funeral Home in Wuhan, "the number of dead bodies our work unit has hauled in the past couple of days has soared." The person claimed to be "hardly able to carry on" and that the funeral home "does not even give us protective clothing." The person also noted that "Wuchang Funeral Home and Hankou Funeral Home cannot hold any more corpses."

Following the Wuhan Bureau of Civil Affairs website, seven funeral homes are in the city. In compiling publicly available data made available by the seven funeral homes, those funeral homes operate about 80 furnaces and cremate about 60,000 corpses each year. For example, Hankou Funeral Home, one of the seven funeral homes, has 14 furnaces available, according to Hong Kong news website Initium Media. However, Hankou Funeral Home claims 30 furnaces, cremates about 20,000 corpses a year, and owns 20 funeral vehicles. If we presume that Hankou has 14 instead of 30 furnaces and carries out cremations for 300 days a year, then the 14 furnaces cremate nearly five corpses a day each. And if we presume that only half of Hankou's 20-funeral vehicle fleet is in use, then each vehicle carries an average of seven corpses to the funeral house every day.

In August 2016, Wuhan Evening News published an article about how cremation technicians have created a so-called "five-step combustion method" to shorten the average time of cremation from 50 to 60 minutes to about 45 minutes. Assuming that actual cremation time is double what the cremation technicians claim (90 minutes) and that only half of the 80 furnaces in Wuhan are working (40 furnaces), then the city can "dispose of" 640 corpses a day via cremation. According to the media interviews I wrote above, this figure is close to the 700 corpses "disposed of" on January 24.

Pursuing publicly available data made available by the seven funeral homes, the funeral homes cremate 60,000 corpses annually on average. Therefore, after accounting for the average death rate of 200 per day (60,000 corpses divided by 300 days), there has been an increase of at least 440 deaths per day in Wuhan since the CCP officially acknowledged the coronavirus outbreak. The increased death rate over the ten days from January 20 to January 29 could be 35.2 times that of the official 170 coronavirus deaths (as of January 29), or 5,984 deaths.

According to a January 30 paper in the British medical journal The Lancet, the paper's authors examined 99 patients infected with COVID-19 between January 1 to

January 20. They found a mortality rate of 11 percent at the cutoff for the research study (January 25). Michael Ryan, executive director of WHO's Health Emergencies Programme, said that the fatality rate of known cases and deaths is around 2 percent. Therefore, in extrapolating from 5,984 deaths between January 20 to January 29, the actual number of confirmed cases could be between 54,400 (11 percent mortality rate) to 299,200 (2 percent mortality rate).

In comparing central government and local government data on the coronavirus, there are obvious signs that one side or all sides are actively partaking in fraudulent behavior.

According to central government data, as of January 29, there are 7,711 confirmed cases (59.5 percent from Hubei) and 170 deaths (95.3 percent from Hubei) in China's 31 provinces, as well as 81,947 people under medical observation (32.5 percent from Hubei). It is striking that while only about 60 percent of confirmed cases in China come from Hubei, over 95 percent of deaths occurred in the province. Because the coronavirus is new, there ought not to be such a glaring discrepancy between the mortality rates in Hubei compared to other regions. There are several plausible reasons for this phenomenon: Other provinces have better and more medical equipment and treatment than Hubei, leading to vastly lower mortality rates; all regions aside from Hubei are hiding their actual coronavirus confirmed cases and mortality rate; or the Hubei local government is hiding the exact number of confirmed cases, leading to an extremely high mortality rate.

Several events and evidence point to and support the accidental laboratory leak of the SARS-CoV-2 virus, but there is no solid evidence that the SARS-CoV-2 is the result of Zoonosis. Yet, no vector species were found to believe that the SARS-CoV-2 jumped from bat to vector to human. (*A vector is the intermediate host of the virus; from which virus transmission to humans is possible*)

V

The Ghost Cousin

"Mystery creates wonder and wonder
is the basis of man's desire"
- Neil Armstrong

The ghost cousin is an exciting title, but when I was studying and researching about RaTG13, I came to know some interesting facts. I could not decide the chapter's name; I was looking to name this chapter RaTG13, but suddenly, I realized some shocking facts, then I decided to call this chapter 'The Ghost Cousin.'

Before we step into a deep understanding of RaTG13, it is essential to understand some basic terms.

What is a genome?

A genome is an organism's complete set of genetic instructions. Each genome contains all of the information needed to build that organism and allow it to grow and develop.

Our bodies are made up of millions of cells (100,000,000,000,000), each with its own complete set of instructions for making us, like a recipe book for the body. This set of instructions is known as our genome and comprises D.N.A.?. For example, each cell in the body, a skin cell or a liver cell, contains this same set of instructions.

The instructions in our genome are made up of D.N.A. Within D.N.A. is a unique chemical code that guides our growth, development, and health. This code is determined by order of the four nucleotide bases that make up D.N.A., adenine, cytosine, guanine, and thymine, A, C, G, and T for short. D.N.A. has a twisted structure in the shape of a double helix. Single strands of D.N.A. are coiled up into structures called chromosomes. Your chromosomes are located in the nucleus within each cell. Within our chromosomes, sections of D.N.A. are "read" together to form genes?. Genes control different characteristics such as eye color and height. All living things have a unique genome. The human genome comprises 3.2 billion bases of D.N.A., but other organisms have different genome sizes.

8th November 2021,

My native place, 11:40 PM

I was eager to see one of my best friends and a brilliant microbiologist Darpan, who is also an owner of a pathology laboratory. After college, he immediately joined a pathology laboratory for an internship. Later, the laboratory owner permanently migrated to the U.S.A. and handed over the laboratory to my friend Darpan. Darpan was brilliant in academics; he completed a Graduation in Microbiology and Masters's in laboratory technology. Therefore, he was the perfect person to discuss the origins of COVID, and I was very sure I would get all the authentic information I needed to understand in writing this book.

We spent long hours discussing socio-political issues and many other topics till late at night at his laboratory. Then, we used to go out for street food and parties! But after March-2020, it all changed.

On the current day [8th November 2021], we decided to discuss the origin of covid-19. As I mentioned earlier, Darpan is a brilliant Microbiologist and Pathologist. I was ready to face brutal counter questions from him. According to him, it is possible that SARS-CoV-2 could have jumped from animals to humans directly or by any vector. Vector means an intermediate animal species between bats and humans.

"During SARS 2002-2004 pandemic, the SARS coronavirus was jumped to humans via an intermediate animal called Chinese wild civets and raccoon dogs! So it is quite possible that this SARS-CoV-2 could have originated like this." Darpan put his point.

"I do not deny this possibility as far as SARS concern. But I do not think this is applicable for SARS-CoV-2." I replied him.

"Any evidence?" he asked me.

"Many a full series of events and the sequence of SARS-CoV-2 itself is considered as the smoking gun!" I replied.

"Interesting, I would like to know about those events," Darpan replied.

"Sure, I will send a copy of my book to you!" I replied

And we continued our discussion for almost there and half hours. I asked him some basic but most important questions about the virus structure and its behavior during the debate. He also handed me a book with vast information about various viruses and information about SARS coronavirus. The book was very heavy, But the knowledge inside was worthy. I am thankful to Darpan; he helped me clear my doubts about writing this book. I believe the truth must be authentic.

Darpan's pathology laboratory is in a densely crowded market. I asked him to suppose any highly contagious species accidentally leaks from his laboratory; what would his first step be to counter this leak and stop its spread? He took a few seconds and replied that he would first inform the emergency services and relative local government bodies.

"Then?" I asked him.

"I would take the corrective control measures to stop the further spread!" He replied.

"Correct! Anything else you want to add?" I asked him.

"No, but I will do whatever I can do to stop this leak and its spread!" He replied.

"I am impressed with your honesty Darpan! But my friend, you forgot to consider the business!" I replied.

"Yes, it will impact on my business but would have a double impact on the reputation too." He replied.

"Exactly! These are some of the reasons to say the COVID-19 pandemic results from Lab. A leak from WIV. Just imagine that if WIV authorities accept that the SARS-CoV-2 virus has been leaked from WIV, what kind of impact will occur on China's reputation. Funding to such laboratories will be banned as well, and international pressure will increase on China." I replied.

"Sounds interesting, and I am also excited to know more about the Lab. Leak." He replied.

To know in-depth about the SARS-CoV-2 virus, we must understand its closest known relative. The closest [96%] similar virus to SARS-CoV-2 is RaTG13.

According to Shi Zhengli and her team of virologists, the SARS-CoV-2 is the closest relative of the RaTG13 Bat coronavirus. She and her team also postulated a short region of RNA-dependent R.N.A. polymerase (RdRp) from a bat coronavirus (BatCoV RaTG13) previously detected in *Rhinolophus affinis* type batsfrom Yunnan province—showed high sequence identity to 2019-nCoV. She and her team carried out a full-length sequencing (*Sequencing means determining the order of the four chemical building blocks - called "bases" - that make up the D.N.A. molecule.*) The sequence tells scientists the genetic information carried in a particular D.N.A. segment. Simplot analysis showed that 2019-nCoV SARS-CoV-2 was highly similar throughout the genome to RaTG13, with an overall genome sequence identity of 96.2%. Phylogenetic analysis of the full-length genome and the gene sequences of RdRp and spike (S) showed that for all sequences—RaTG13 is the closest relative of 2019-nCoV, and they form a distinct lineage from other SARSr-CoVs. A Spike protein is a Glycoprotein that protrudes from the envelope of a virus-like Coronavirus and facilitates the entry of the virion (a complete virus particle that consists of an R.N.A. or D.N.A. core with a protein coat sometimes with external envelopes and that is the extracellular infectious form of a virus) into the host cell by binding to a receptor on the surface of the host cell followed by the fusion of the viral and host cell membranes. Like a key in a lock, these spike proteins fuse to receptors on the surface of cells, allowing the virus's genetic code to invade the host cell, take over its machinery and replicate.

The S genes of 2019-nCoV and RaTG13 are longer than other SARSr-CoVs. Shi Zhengli further adds to her study that ACE2 is the known receptor for the SARS-CoV. Her research (as she mentions) to determine whether 2019-nCoV also uses ACE2 as a cellular entry receptor conducted virus infectivity studies using HeLa cells that expressed or did not express ACE2 proteins from humans, Chinese horseshoe bats, civets, pigs, and mice. We show that 2019-nCoV can use all ACE2 proteins, except for mouse ACE2, as an entry receptor to enter ACE2 expressing cells, but not cells that did not express ACE2, indicating that ACE2 is probably the cell receptor through which 2019-nCoV enters cells. Interestingly Shi mentions in her article on 'Nature.com' that 'It appears that the virus is becoming more transmissible between humans. We should closely monitor whether the virus continues to ***evolve*** to become more virulent.' The surprising word in her statement is 'evolve'; the evolution of a living species takes decades!

In counter of Shi Zhengli's article on Nature.com, an article authored by Xiaoxu Sean Lin, Global Health Knowledge Exchange Inc., Silver Spring, MD, U.S.A. Shizhong Chen, Institute, San Diego, CA, U.S.A opened up the discussion and questioned the quality and credibility of the article and the methods used to identify the RaTG13.

After the COVID-19 outbreak started in China, one group of Chinese scientists, including Peng Zhou and Zhengli Shi from the Wuhan Institute of Virology (WIV), published an impactful article in Nature on 3rd February 20201. The paper was first submitted on 20th January 2020. It claimed that full-length coronavirus genome sequences were obtained from five patients, and the viral genome is 96.2% identical at the whole-genome level to a bat coronavirus RaTG13 strain. This paper has been highly cited since its publication. The only one presented this unique bat coronavirus strain, which supports a potential viral zoonotic transmission from bat to human. However, after careful reading into this paper, the origin, identification, and characterization of the BatCoV RaTG13 strain emerge as outstanding questions. In addition, some experimental methodology, data quality, and experimental procedures described in this paper are concerning and warrant further validation. The mysterious origin of bat coronavirus strain

RaTG13 Shi's Nature paper1 stated that "We then found that a short region of RNA-dependent R.N.A. polymerase (RdRp) from a bat coronavirus (BatCoV RaTG13)—which was previously detected in Rhinolophus affinis from Yunnan province—showed high sequence identity to 2019-nCoV. We carried out full-length sequencing on this R.N.A. sample (GISAID accession number EPI_ISL_402131). Simplot analysis showed that 2019-nCoV was highly similar throughout the genome to RaTG13, with an overall genome sequence identity of 96.2%."

According to the information on GISAID regarding RaTG13, this bat coronavirus strain was collected in July 2013, nearly seven years ago. Zhengli Shi's previous publications related to bat coronaviruses have identified 365 bat coronavirus strains from Yunnan province, from a total of 1981 bat samples2-6. However, despite high-profile publications describing single coronavirus discoveries, all these publications did not mention this unique strain RaTG13. In addition, Shi's previous study5,6 of BatCoV RdRp in 2016 did not report about this RaTG13

strain, but highlighted another bat coronavirus strain, BtCoV/4991, also identified in the same bat species of *Rhinolophus affinis.* What is most unusual is that the short region of the RdRp gene used to distinguish different lineages of bat coronaviruses in the phylogenetic analysis5 showed 100% nucleotide identity between BtCoV/4991 and RaTG137. This raises the serious question of whether RaTG13 and BtCoV/4991 are the same strain, as this 100% identity is not at the amino acid level but at the nucleotide level. Suppose these two were indeed two separate strains of bat coronaviruses. In that case, Shi's group should also report, or even first find out, that BtCoV/4991 shows high similarity with SARS-CoV-2 RdRp, as BtCoV/4991 RdRp sequence was previously sequenced and submitted to GenBank (Accession number KP876546) in 2016. And if they are the same strain, what was the rationale to designate two separate names to the same thing?

In addition, through GenBank blast analysis, we found that BtCoV/4991 partial RdRp gene sequence has 98.65-98.92% nucleotide homology with that of SARS-CoV-2 strains, but only 87% homology with two other bat SARS-like coronavirus strains (bat-SL-CoVZXC21, GenBank: MG772934.1; bat-SL-CoVZC45, GenBank: MG772933.1) that have relatively high genome homology (89%) with SARS-CoV-2. Therefore, this unique feature of very high homology in the RdRp gene should warrant more studies on the BtCoV/4991 strain, yet this strain was not mentioned in this Nature paper.

Currently, RaTG13 is the only BatCoV that shows as high as 96% entire genome homology with SARS-CoV-2, as BtCoV/4991 entire genome sequence was not available. The sequence of RaTG13 was a total outlier in the Phylogenetic analysis when compared to other bat coronaviruses. Jiu meng Sun et al. found that in the maximum likelihood Phylogenetic analysis a middle segment of the SARS-CoV-2 genome (from nt 13522 to nt 23686) and RaTG13 does not cluster with Sarbecovirus, a subgenus of the beta-corona virus that bat SARS-like coronaviruses belong to. Lam et al. found six pangolin coronavirus sequences with 85.5% to 92.4% homology to SARS-CoV-2, less than the 96% homology that RaTG13 has. However, the pangolin coronavirus spike protein sequences shared five key amino acids in the receptor-binding domain (RBD) with SARS-CoV-2, while RaTG13 only shared one key amino acid in its RBD with SARS-CoV-2.

Meanwhile, only SARS-CoV-2 has a G/C rich polybasic furin-cleavage site at the S1/S2 junction in the spike protein, while RaTG13 or pangolin coronaviruses do not have this cleavage site. No mention of this polybasic cleavage site was reported in Shi's Nature paper, despite it being a major feature that differentiates SARS-CoV-2 from RaTG13 spike proteins. Therefore, these unique features of the RaTG13 strain sequence make the SARS-CoV-2 origin and immediate animal reservoir issues deeply compounded.

As Shi's Lab's routine task to study bat coronavirus spike proteins and their RBD domains, it is odd that the Shi group allegedly did not pursue RaTG13 or BatCoV/4991 for nearly seven years to characterize their S proteins further. In 2013, her group published an article in Nature describing the discovery of two bat coronaviruses, Rs3367, and SHC014 that share considerable sequence similarities in the RBD region with SARS2. However, while studies with RBD motifs in Rs3367 and SHC014 made breakthroughs in coronavirus research, it is rather unusual that RaTG13 with unique features failed to trigger any interest within the Shi group so far. Moreover, even in the most recent 2020 publication10 by Shi's group, SHC014 and other bat SARS-like coronaviruses, but not RaTG13 or BatCoV/4991, were used to study the interaction among S proteins and bat ACE2 variants while making pseudoviruses with the S protein from RaTG13 or BatCoV/4991 would not be challenging for her Lab.

The only functional characterization experiment related to RaTG13 was a structured study using the synthetic S protein based on the RaTG13 S gene sequence in the GenBank (accession number QHR63300.2), not using any RaTG13 virus sample. In addition, the affinity of S protein from RaTG13 with human ACE2 needs to be characterized as well, since the affinity of human ACE2 is 4.5 times higher for SARS-CoV-2 than for SARS-CoV12. If the affinity of RaTG13 S protein is similar to SARS CoV-2, this suggests human transmission is readily possible from a RaTG13 "sister virus" (or progenitor virus of SARS-CoV-2). On the other hand, if RaTG13 S protein affinity is much lower, this suggests significant adaptation could have been required between the progenitor virus and SARS-CoV-2 to gain tight human ACE2 interaction capacity.

Why was The Genome Sequence not published Despite having unique S protein properties? It is impossible to ignore such a remarkable Novel Corona Virus as Shi Zhengli, and her team claimed they had found it in 2013. It also raises the thought of any secret bio-weapon program of China! Can we neglect this possibility? I do not think we can ignore it. There is evidence that top Chinese military personnel has visited WIV multiple times. We will focus on that area later.

In essence, only the genome sequence of RaTG13 has been made available so far. Key information related to identifying and isolating this RaTG13 virus strain is missing in Shi's Nature paper1. A series of important questions regarding RaTG13 remains to be answered:

What bat tissue/organ samples were collected in 2013 and then subjected to viral isolation or sequencing to obtain this BatCoV RaTG13?

Did RaTG13 cause diseases in this bat species of Rhinolophus affinis? Was the bat sample for RaTG13 collected in the same cave as BatCoV/4991?

In addition, this Nature paper1 did not mention any efforts to rule out the possibilities that the RaTG13-related bat sample collected in 2013 might have been mixed with other bat samples, or the bat was infected with two different strains of coronaviruses, as co-infections with two or more strains of coronaviruses in bats was not a rare event. Therefore, if the current RaTG13 sequence identified was indeed a mixed sequence of two strains of bat coronavirus due to random sequencing of bat samples without prior isolation of the virus, then all the current Phylogenetic analyses involving RaTG13 were futile and subjected to correction. And at the time you are reading this book, they have not claimed such a thing yet, and if they do in the future, they will lose their remaining credibility!

Considering the above-mentioned strange features in the RaTG13 genome and the fact that no prior or recent studies have been conducted using live viral stocks of this virus, concerns regarding the history and existence of the BatCoV RaTG13 strain are reasonable and legitimate. This is an important issue because the existence of this unique BatCoV RaTG13 is significantly involved in the analyses of the evolutionary relationship among SARS-CoV-2, bat SARS-like coronaviruses, and other pangolin coronaviruses. If the authors had collected these two specimens (for RaTG13 & BatCoV/4991) that were outliers to other bat coronaviruses in one location in Yunnan, it would suggest that maybe other closely related sister viruses also exist in the same region. This is critical to know as these SARS-CoV-2 "sister viruses" could pose a significant threat to another global pandemic and provide key information on the evolutionary origin of the SARS-CoV-2 virus. Furthermore, additional virus sequences in the SARS-CoV-2, RaTG13, or BatCoV/4991 sub-lineage would aid understanding of the origin of the insertion between S1 and S2 of a furin cleavage site in SARS-CoV-2 that is associated with increased Pathogenicity. Under the openness and ethics guidelines for scientific publications, particularly in

Nature and given the magnitude of the pandemic's impact, the Shi's team must provide samples of RaTG13 & BatCoV/ 4991 for other scientists to conduct independent verification experiments and further characterization of this RaTG13 or BatCoV/4991 virus strain.

Shi's Nature paper1 mentioned that "Of the 10,038,758 total reads—of which 1,582 total reads were retained after filtering of reads from the human genome—1,378 (87.1%) sequences matched the sequence of SARSr-CoV (Fig. 1a). By de novo assembly and targeted PCR, we obtained a 29,891base-pair CoV genome that shared 79.6% sequence identity to SARS-CoV BJ01 (GenBank accession number AY278488.2). High genome coverage was obtained by remapping the total reads to this genome (Extended Data Fig. 1). This sequence has been submitted to GISAID (https://www.gisaid.org/) (accession number EPI_ISL_402124)." However, the methods described here to obtain full SARS-CoV-2 sequence have major flaws. First, characterization of novel viruses from patient samples using next-generation sequencing (NGS) technology must overcome the challenges posed by the high degree of genetic diversity observed across most virus families, especially for R.N.A. viruses. Due to the error rate in R.N.A. viral replication, what existed in a patient sample are usually viral quasispecies or mixed populations. Therefore, the method of random sequencing plus de novo assembly used in this study should only be used as the initial characterization. What is needed is to redo the NGS using the isolated virus stock so that the volume of raw reads related to the target sequence would be significantly enhanced15. Then, reference assembly (using a SARS-CoV strain or Bat SARS-like CoV strain as reference) can be applied to obtain comprehensive coverage of the full viral genome with ample depth. Therefore, this study uses only 1,378 reads from random amplification to conduct de novo assembly and get near-complete genome coverage (as shown in Extended Data Fig.1) for such a large R.N.A. virus genome (near 30K in total length) is beyond a miracle.

Meanwhile, for regions with a high chance of mutations such as spike protein open reading frame, very in-depth coverage of raw reads is often needed to ensure the accuracy of sequencing data. Random sequencing from patient samples would not work for genome regions with high variations, yet the paper did not mention any extra efforts to address such concerns. The accuracy of the full genome sequences obtained in this study should be seriously challenged.

In the "Extended Data Fig. 6: Isolation and antigenic characterization of 2019-nCoV" of this paper published on Nature.com by Shi Zhengli, it did mention carrying out "metagenomics analysis of supernatants from Vero E6 cell cultures". Then, if NGS was conducted using viral supernatants from cell culture, the authors need to explain why those sets of NGS data were not presented in the paper. Still, instead, NGS data from random sequencing of patient samples was presented. Therefore, this study's SARS-CoV-2 genome sequence submitted to the GISAID (accession number EPI_ISL_402124) needs to be verified for its quality.

Shi's Nature paper1 mentioned that "four more full-length genome sequences of 2019-nCoV (WIV02, WIV05, WIV06, and WIV07) (GISAID accession numbers EPI_ISL_402127–402130) that were more than 99.9% identical to each other were subsequently obtained from four additional patients using next-generation sequencing and PCR (Extended Data Table 2)." However, since these full-length genome sequences were obtained without isolating the viruses from the patient samples (as explained in the footnote for Table 2)1, the quality of the genome sequences could be compromised. Therefore, to ensure the quality of data submitted to GISAID, the Zhengli Shi team should provide raw NGS reads to open platforms so that other scientists could review and re-analyze the raw sequencing data related to these important COVID-19 patient samples.

On the official website of Wuhan Institute of Virology, the Institute leadership published an open letter to all its staff and graduate students on 17th February 2020. This open letter stated that the SARS-CoV-2 strain was isolated on 5th January 2020, and its full genome sequence was obtained as early as 2nd January 202016. However, regarding the procedures for viral isolation, Nature's paper also described in the Method section that "the culture supernatant was examined for the presence of virus by R.T.–PCR methods developed in this study." This indicated that the viral isolation could NOT be initiated simultaneously with genome sequencing experiments from patient samples because the qRT-PCR experiments would need specific primers and probes that could only be designed and produced after the genome sequencing was completed. This would suggest that Shi's team only had as little as 2-3 days to obtain the viral isolate. Therefore, the experimental procedures described in this paper were surely rushed and need further validation.

Play, Pause, and Rewind

Play, pause, and rewind; Play, pause, rewind as I was eagerly tapping my laptop's trackpad, my colleagues were weirdly looking at me. They were not happy with my behavior with my laptop, but at the same time, they were also wanted to know what am I doing! So I continued for almost one and half an hour. The video was an interview of

Wang Yanyi, Director of Wuhan Institute of Virology, to the CGTN channel.

When CGTN reporter Hu Chao asked her about the SARS-CoV-2 virus, she denied having any live SARS virus at WIV. Next, Hu Chao asked her about the possible laboratory leak from the Wuhan Institute of virology; Wang Yanyi first explained that it is a pure fabrication. Further, she adds that they were sequencing the SARS-CoV-2 virus; no live RaTG13 virus was present at WIV. Finally, Hu Chao asked her about the published thesis in 2018 at periodical Nature about the novel coronavirus originating from bats. A surprising and ridiculous answer to this question was that the virus mentioned in the 2018 thesis was not the SARS-CoV-2 that caused the pandemic. She also added that the virus mainly causes diarrhea and death among piglets later; this virus is known as SADS! The similarity of genome sequences between SADS and SARS-CoV-2 is only 50%, a big difference.

Her disclaimer having no live RaTG13 coronavirus at WIV is indigestible! She also claims that scientists at WIV neither isolated nor obtained RaTG13, but they have the Genome Sequence of RaTG13.

In summary, as a study isolated the first SARS-CoV-2 strain and identified a bat coronavirus with high homology to SARS-CoV-2, all data relating to viral genomes must be of top quality since many studies used or might use them as reference sequences. Meanwhile, although the leadership in Wuhan Institute of Virology might highlight their impressive speed to complete all related experiments described in the paper16, the accurate patient sample collection date and sequencing data with better quality needs to be recorded in the scientific paper. In addition, there have been no studies on RaTG13's infectivity in bat/

human cells or animal models, its interactions with antibodies or antiviral drugs. There is a lack of understandings of RaTG13's virulence, transmissibility, Pathogenicity, immune epitopes, immune evasion mechanism, etc. This was because WIV did not isolate the RaTG13 virus and did not have any related viral stocks if the statement from Dr. Yanyi Wang (the director of WIV) in a recent T.V. interview was accurate. Unfortunately, the video of that interview is no more available now!

Therefore, a careful examination of the related RaTG13 samples and raw data sets of its genome sequencing is warranted to exclude any possible errors or the potential co-infection of two different strains of coronavirus. And the authors need to clearly explain the relationship between RaTG13 and BatCoV/4991, whether they were the same strain or two closely related strains. This paper was rushed to make a premature connection between bat coronavirus and SARS-CoV-2, drawing a potential bat origin scenario to support SARS-CoV-2 zoonotic transmission from bat to human. However, this connection was based on a potential bat coronavirus strain RaTG13 that may not truly exist, considering its key information missing: such as no related bat sample description, no sequencing procedure details published, confusion/identity issue with BatCoV/4991 strain, unusual sequence features, no viral isolation, and related characterization, et al.

In the focus of these concerns, we call for the retraction of this Nature paper to further verify the sequencing data, patient sample collection date and provide more information regarding the origin, identification, and characterization of this BatCoV RaTG13. Proper verification should involve Dr. Zhengli Shi sending the RaTG13 and BatCoV/4991-related bat samples to other non-collaborating laboratories to be analyzed independently. And this Nature paper1 should be cautious on making the "probable bat origin" hypothesis before RaTG13 existence could be confirmed.

From, above all facts and data, some questions are still alive! According to Dr. Shi Zhengli, she and her team had already derived the genome sequence of RaTG13 earlier in 2013, but they never published it or submitted the sequence in Gene bank! And on the other hand, if we trust the on-camera statement of Dr. Yanyi Wang, RaTG13 was neither physically isolated nor obtained!

For an experiment to be considered valid, it must be repeated independently in other laboratories. And to conduct such experiments, the live sample of RaTG13 is necessarily available, which is not possible as per the statement of the director of WIV, Dr. Yanyi Wang.

A research paper published by Yue Li1, XinaiYang, Na Wang, Haiyan Wang, Bin Yin, Xiaoping Yang & Wenqing Jiang from Department of Respiratory Diseases, Qingdao Hai ci Hospital, PR China under the title **'The divergence between SARS-CoV-2 and RaTG13 might be overestimated due to the extensive R.N.A. modification'** The motive of the experiment to understand the phylogeny, divergence, and origin of SARS-CoV-2.

Why RaTG13 is a Ghost Cousin of SARS-CoV-2?

A preprint published by Lin X and Chen S paper presented a premature hypothesis of the potential bat origin of SARS-CoV-2 while the RaTG13 strain was not successfully isolated. The authors also pointed out the methodology, data quality, and experiment procedures described in the paper published in Nature2. Lin X and Chen S call for the authors to provide additional data, to share related samples to be verified and further characterized by other scientists.

Anonymous scientist (NerdHasPower) published his opinion about the RaTG13 virus and his doubts about the validity of the data related to that virus.

"The sequence of RaTG13 was reported by Zhengli Shi, a researcher of the Wuhan Institute of Virology and the biosafety level 4 (P4) laboratory for virology research. Shi ZhengLi published a paper in Nature. She compared the freshly obtained sequence of the coronavirus with those of other coronaviruses and thus delineated an evolutionary path of this new virus. In this publication, suddenly and out of nowhere, Shi reported this bat coronavirus, RaTG13, which pampered the public and seemingly helped shape a consensus in the field that COVID-19 is of a natural origin. As stated in the paper, RaTG13 was discovered from Yunnan province, China, in 2013. According to credible sources, Shi has admitted to several individuals in the field that she does not have a physical copy of this RaTG13 virus. Her Lab allegedly collected some bat feces in 2013 and analyzed these samples for the possible presence of coronaviruses based on genetic evidence. To put it into plainer words, she has no physical proof for the existence of this RaTG13 virus. She only has its sequence information, which is nothing but a string of letters alternating between A, T, G, and C."

RaTG13, if it Physically exists, should never be neglected by Shi for seven years

The sequence of RaTG13 is highly alarming – it clearly shows the potential of the virus to infect humans.

Within the spike protein of a β coronavirus, a critical piece called receptor-binding domain (RBD) dictates whether or not this virus can use the ACE2 receptor on the surface of our cells and thereby infect humans. As a routine, when

Shi's team finishes collecting samples and confirms the presence of a coronavirus, they would first look at the sequence of the virus' RBD. If there is a resemblance between this sequence and that of the SARS virus (rarely so), their blood will boil because they have found something that may jump over to humans. It also means that top-journal publications are coming their way.

In 2013, Shi became famous in the coronavirus field by publishing in Nature two bat coronaviruses (Rs3367 and SHC014), which share considerable sequence similarity with SARS in the RBD region (2). For the first time, this work proved a bat origin of SARS. In the following years, her team published articles featuring additional bat coronaviruses that share these important sequence motifs (3, 4).

What does an RBD sequence look like? Figure 1 is the sequence comparison between SARS RBD and the RBDs of the bat coronaviruses that Zhengli Shi published in high-profile journals (2-4). Compared to SARS (top), many bat coronaviruses (most of the ones in the bottom half) had substantial deletions in their RBDs and are thus likely defective in targeting humans. In contrast, some bat coronaviruses (upper half) resemble SARS in the completeness of the RBD sequences and contain amino acids similar to their SARS counterparts at some of the five locations known to be critical for binding human ACE2 receptors. With these dazzling features, the field perceived this group of viruses as breakthroughs.

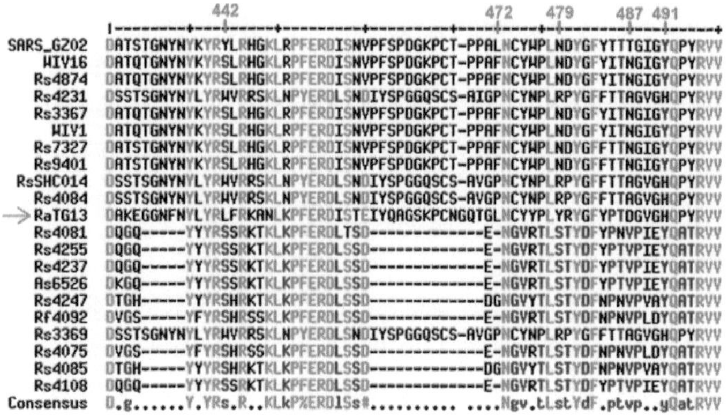

```
                  442                            472  479    487 491
          |--------+---+-----+---------+---------+----+----+-+-------+---+-----+
SARS_GZ02 DATSTGNYNYKYRYLRHGKLRPFERDISNVPFSPDGKPCT-PPALNCYWPLNDYGFYTTTGIGYQPYRVV
    WIV16 DATQTGNYNYKYRSLRHGKLRPFERDISNVPFSPDGKPCT-PPAFNCYWPLNDYGFYITNGIGYQPYRVV
   Rs4874 DATQTGNYNYKYRSLRHGKLRPFERDISNVPFSPDGKPCT-PPAFNCYWPLNDYGFYITNGIGYQPYRVV
   Rs4231 DSSTSGNYNYLYRWVRRSKLNPYERDLSNDIYSPGGQSCS-AIGPNCYNPLRPYGFFTTAGVGHQPYRVV
   Rs3367 DATQTGNYNYKYRSLRHGKLRPFERDISNVPFSPDGKPCT-PPAFNCYWPLNDYGFYITNGIGYQPYRVV
     WIV1 DATQTGNYNYKYRSLRHGKLRPFERDISNVPFSPDGKPCT-PPAFNCYWPLNDYGFYITNGIGYQPYRVV
   Rs7327 DATSTGNYNYKYRSLRHGKLRPFERDISNVPFSPDGKPCT-PPAFNCYWPLNDYGFFTTNGIGYQPYRVV
   Rs9401 DATSTGNYNYKYRSLRHGKLRPFERDISNVPFSPDGKPCT-PPAFNCYWPLNDYGFFTTNGIGYQPYRVV
 RsSHC014 DSSTSGNYNYLYRWVRRSKLNPYERDLSNDIYSPGGQSCS-AVGPNCYNPLRPYGFFTTAGVGHQPYRVV
   Rs4084 DSSTSGNYNYLYRWVRRSKLNPYERDLSNDIYSPGGQSCS-AVGPNCYNPLRPYGFFTTAGVGHQPYRVV
-> RaTG13 DAKEGGNFNYLYRLFRKANLKPFERDISTEIYQAGSKPCNGQTGLNCYYPLYRYGFYPTDGVGHQPYRVV
   Rs4081 DQGQ-----YYYRSSRKTKLKPFERDLTSD-------------E-NGVRTLSTYDFYPNVPIEYQATRVV
   Rs4255 DQGQ-----YYYRSSRKTKLKPFERDLSSD-------------E-NGVRTLSTYDFYPTVPIEYQATRVV
   Rs4237 DQGQ-----YYYRSSRKTKLKPFERDLSSD-------------E-NGVRTLSTYDFYPTVPIEYQATRVV
   As6526 DKGQ-----YYYRSSRKTKLKPFERDLSSD-------------E-NGVRTLSTYDFYPTVPIEYQATRVV
   Rs4247 DTGH-----YYYRSHRKTKLKPFERDLSSD-------------DGNGVYTLSTYDFNPNVPVAYQATRVV
   Rf4092 DVGS-----YFYRSHRSSKLKPFERDLSSD-------------E-NGVRTLSTYDFNPNVPLDYQATRVV
   Rs3369 DSSTSGNYNYLYRWVRRSKLNPYERDLSNDIYSPGGQSCS-AVGPNCYNPLRPYGFFTTAGVGHQPYRVV
   Rs4075 DVGS-----YFYRSHRSSKLKPFERDLSSD-------------E-NGVRTLSTYDFNPNVPLDYQATRVV
   Rs4085 DTGH-----YYYRSHRKTKLKPFERDLSSD-------------DGNGVYTLSTYDFNPNVPVAYQATRVV
   Rs4108 DQGQ-----YYYRSSRKTKLKPFERDLSSD-------------E-NGVRTLSTYDFYPTVPIEYQATRVV
Consensus D.g......Y.YRs.R..KLkP%ERDlSs#.......... ....Ngv.tLstYdF.ptvp..yQatRVV
```

Sequence alignment comparing the RBDs of SARS (top) and RaTG13 (red arrow) to RBDs of bat coronaviruses that Zhengli Shi published in high-profile journals from 2013-2017 (2-4). Amino acid residues highlighted by Shi as critical for binding human ACE2 receptor (2) are labeled in red text on top. Alignment was done using the MultAlin webserver (http://multalin.toulouse.inra.fr/multalin/). [Image Credits: https://nerdhaspower.weebly.com/ ratg13-is-fake.html]

"That is why it is equally important to resolve the mystery of 'The Ghost Cousin' RaTG13 as to resolve the mystery of the origin of SARS-CoV-2!"

The hunt for RaTG13's physical evidence if any present, will continue. It will be like following and tracking a ghost.

Now, you might understand why I have given the name 'The Ghost Cousin' to RaTG13!

The chapter on RaTG13 ends here, but the hunt for smoking guns will continue for both RaTG13 and SARS-CoV-2!

VI
The Experiment

September 2021

It has been more than three months of writing this book. In these three months, many things had been changed. At the end of August, we were finally free to go home and stay with our family. My family and I were fully vaccinated. The Indian Government and Medical professionals showed the world what India could do. Every Indian and I was proud that India vaccinated their citizens and helped multiple countries worldwide with a supply of vaccines.

My friend Kalpesh Ji was also leaving the company and ready to start a new assignment at a much better position in another company. He had resigned in mid-September – 2021 before his contract ended with his company. I wished him the best of luck in the future, and we promised each other to stay in touch! I told him that I would send him a copy of this once It gets published.

November 10, 2021

At Darpan's Pathology Laboratory

"Darpan, I feel it is not reasonable to point out towards someone or a laboratory holding responsible for the pandemic without strong evidence," I said to Darpan.

"I am surprised that you are telling this!" He looked at me and said.

"Yes, I internally feel that something must be solid to say that the laboratory could be the only source for the SARS-CoV-2 outbreak." I replied to him, opening my diary and noting down 'solid evidence.'

"The truth must be balanced, not biased!" He replied.

"We know that the SARS-CoV pandemic as a result of Zoonosis, and shreds of evidence like physical evidence are also pointed towards the Civets sold in Guangdong market. Similarly, SARS-CoV-2 could also result from animal to human transmission. But all the circumstantial evidence and events only point towards the WIV. This is clear that the Chinese government would not allow foreign bodies to investigate inside the laboratory. Moreover, none of the laboratory scientists or admin person will speak about it, or else they would be wiped off like other whistleblowers. So the only way to find out what could have happened inside WIV is to connect dots of reports, documents, and published researches." So I said to Darpan by sipping my soft drink.

"The RaTG13 is the closest cousin of SARS-CoV-2 the match between them is 96.4% according to a well-known virologist Edward Holmes for a RaTG13 it would take about 50 years into a SARS-CoV-2. His statement is on camera.

He and David Relman, Professor of Microbiology and Medicine, Stanford is also on the side of natural origin. He said nearly all previous outbreaks of emerging viruses and pathogens had been natural, and he is right. Also, another Professor of Microbiology at Tulane Medical School, New Orleans, Robert Garry, said that the virus could not naturally evolve in the dying miners in 2013 because that would take decades for natural evolution." Darpan said to me.

"Is there any other way that the RaTG13 could evolve to SARS-CoV-2 faster?" I asked him.

"Yes, that is possible in a high-tech virology laboratory having BSL-4 kind of safety and facility! It is called genetic engineering in technical language, and the research or that experiment is called Gain-of-Function." Darpan replied with excitement.

"So, you want to say that like Hollywood movies, you can program a virus to transmit among humans!" I asked Darpan in a light mood.

"Just something like that!" He replied.

The conversation was getting interesting between us as we shared our views on it openly without giving any biased statement. We were considering every possibility from Zoonosis to deliberate leak. However, we both do not want to reach any conclusion in a hurry; even if we found something solid and presentable supporting Zoonosis, we would have considered it.

An article in Independent Science News by Jonathan Latham and Allison Wilson discusses another mechanism, described by Nikolai Petrovsky of Flinders University in Australia, that could have resulted in the SARS-CoV-2 virus that produced the pandemic:

Take a bat coronavirus that is not infectious to humans, and force its selection by culturing it with cells that express human ACE2 receptor, such cells having been created many years ago to culture SARS coronaviruses. Then, you can force the bat virus to adapt to infect human cells via mutations in its spike protein, which would have the effect of increasing the strength of its binding to human ACE2, and inevitably reducing the power of its binding to bat ACE2.

Viruses in prolonged culture will also develop other random mutations that do not affect their function. The result of these experiments is a highly virulent virus in humans but is sufficiently different that it no longer resembles the original bat virus. Because the mutations are acquired randomly by the selection, there is no sign of a human gene jockey, but this is a virus still created by human intervention.

No Pre-Pandemic SRS-CoV-2 Antibodies found

One more convincing reason to believe SARS1-CoV emerged naturally during 2002-2004 is the antibodies. There is evidence that the animal vendors not diagnosed with SARS1-CoV had antibodies to SARS1-CoV. This indicated that the SARS1 virus infected the vendors on previous dates; SARS1 somehow circulated among animals and humans.

No similar evidence has been found yet that could give weight to the natural emergence of SARS-CoV-2 (COVID-19) was emerged naturally. Researchers could discover antibodies from the blood collected in the blood banks or samples collected before the pandemic. This will be substantial evidence that the virus circulated among the animals and periodically infected humans. No such antibodies were found from blood bank samples.

In contrast, to consider that SARS-CoV-2 naturally emerged without pre-pandemic antibodies, we must assume that the SARS-CoV-2 was introduced only once. That single introduction sparked the pandemic, which is impossible to match the set conditions for the natural emergence of the virus.

Gain of Function Research

Gain-of-function (GOF) research involves experimentation that aims or is expected to (and/or, perhaps, actually does) increase pathogens' transmissibility and/or virulence. Such research, when conducted by responsible scientists, usually aims to improve the understanding of disease-causing agents, their interaction with human hosts, and/or their potential to cause pandemics. The ultimate objective of such research is to inform better public health and preparedness efforts and/or the development of medical countermeasures. Despite these essential potential benefits, GOF research (GOFR) can pose risks regarding bio-security and bio-safety.

Gain-of-function research (GoF research or GoFR) is medical research that genetically alters an organism to enhance the biological functions of gene products. This may include an altered pathogenesis, transmissibility, or host range, i.e., the types of hosts that a microorganism can infect. This research intends to reveal targets to predict emerging infectious diseases better and develop vaccines and therapeutics.

In 2014 the administration of U.S. President Barack Obama called for a "pause" on funding (and relevant research with existing U.S. Government funding) of GOF experiments involving influenza, SARS, and MERS viruses in particular. With the announcement of this pause, the U.S. Government launched a "deliberative process" regarding risks and benefits of GOFR to inform future funding decisions—and the U.S. National Science Advisory Board for Biosecurity (NSABB) was tasked with making recommendations to the U.S. Government on this matter.

The reason for that "Pause" was Following the controversy surrounding the research, published in 2012, that led to the creation of highly pathogenic H5N1 (avian) influenza virus strains that were airborne transmissible between ferrets—and more recent reports of bio-safety mishaps involving anthrax, smallpox, and H5N1 in government laboratories.

GOFR is a subset of "dual-use research"—i.e., research that can be used for both beneficial and malevolent purposes' Dual-use research of concern' (DURC) refers to dual-use research for which the consequences of malevolent use would be exceptionally severe (whereas almost any research might be considered "dual-use" broadly conceived—because almost any research, or just about anything for that matter, can be used for some malevolent purpose or other). Of particular concern in the context of life science research is that advances in biotechnology may enable the development and use of a new generation of biological weapons of mass destruction.

The most controversial dual-use life science experiments to date involved the creation of highly pathogenic H5N1 (avian) influenza virus strains that were airborne transmissible between ferrets, which provide the best model for influenza in humans. This research addressed an essential scientific question—i.e., Might it be possible for H5N1 to naturally evolve into a human-to-human transmissible strain and thus result in a pandemic?—and (purportedly) yielded an affirmative answer. After the U.S. National Science Advisory Board for Biosecurity (NSABB) initially recommended that these studies be published in a redacted form (i.e., including key findings, while omitting detailed descriptions of materials and methods). It later approved the publication of revised versions in full, and the papers were published in 2012. Advocates of these studies/publications argued that they would improve surveillance of H5N1 in nature (facilitating early identification of, and thus better respond to, the emergence of potential pandemic strains) and facilitate the development of vaccines that might be needed to protect against pandemic strains of the virus. Critics questioned the validity of claims about such benefits. They argued that the studies might facilitate the creation of biological weapons agents that could kill millions, or possibly even billions, of people.

In layman's language, the Gain of Function engineering involves the artificial modification of an animal virus in a laboratory. In this Gain of Function engineering, a virus is considered a potential threat to humans; all the viruses do not infect and spread among humans. That virus was initially unable to create a pandemic but insert genetic features in that virus; so it can effectively spread among humans and enabled to create a pandemic.

In one line, Gain of Function research enables a natural virus [which cannot infect/spread humans] to spread among humans. The motive to perform such a risky experiment is to understand and prevent future epidemics.

I called Darpan and informed next day,

"I found there is a fair number of research papers available that proves that Shi ZhengLi (Bat Woman), the lead virologist, and her team were researching Bat coronavirus. including the modified virus testing in a mouse-adapted SARS-CoV backbone."

"Now you have the base!" He said.

Shi Zhengli and her team have been working on SARS-CoV since 2013 and much before. Shi Zhengli and her team examine the disease potential of a SARS-like virus, SHC014-CoV, circulating through horseshoe bats. Using the reverse genetics system, Shi Zhengli and her team generated and characterized a chimeric virus expressing the spike of bat coronavirus SHC014 in a mouse-adapted SARS-CoV backbone.

The results indicate that group viruses encoding the SHC014 spike in a wild-type backbone can efficiently use multiple orthologs of the SARS receptor human angiotensin-converting enzyme II (ACE2) and replicate efficiently in primary human airway cells, and achieve in vitro titers equivalent to epidemic strains of SARS-CoV. Additionally, in vivo experiments demonstrate replication of the chimeric virus in mouse lungs with notable pathogenesis. Evaluation of available SARS-based immune-therapeutic and prophylactic modalities revealed poor efficacy; both monoclonal antibody and vaccine approach failed to neutralize and protect from infection with CoVs using the novel spike protein. Based on these findings, they synthetically re-derived an infectious full-length SHC014 recombinant virus and demonstrated robust viral replication both in vitro and in vivo. Their work suggests a potential risk of SARS-CoV re-emergence from viruses currently circulating in bat populations.

In an elaborate article in the Bulletin of the Atomic Scientists, Wade said that much of the work of Chinese virologists on gain-of-function in coronaviruses was performed at the BSL2 safety level lab, which requires taking fairly minimal safety precautions. The pandemic broke out in the Chinese city of Wuhan in December 2019.

"For the lab escape scenario, a Wuhan origin for the virus is a no-brainer. Wuhan is home to China's leading center of coronavirus research, where researchers were genetically engineering bat coronaviruses to attack human cells. They were doing so under the minimal safety conditions of a BSL2 lab. If a virus with the unexpected infectiousness of SARS2 had been generated there, its escape would be no surprise," he said.

As seen in the chapter 'The Epicenter,' the Understanding The Risk of Bat Coronavirus Emergence project was launched in 2014. The leader of the project was Peter Daszak. The project was launched with three particular goals as below,

1. Characterize the diversity and distribution of high spillover-risk SARSr-CoVs in bats in southern China. We will use phylogeographic and viral discovery curve analyses to target additional bat sample collection and molecular CoV screening to fill in gaps in our previous sampling and fully characterize natural SARSr-CoV diversity in southern China. In addition, we will sequence receptor binding domains (spike proteins) to identify viruses with the highest potential for spillover, which we will include in our experimental investigations (Aim 3).
2. Community and clinic-based syndromic surveillance to capture SARSr-CoV spillover, routes of exposure, and potential public health consequences. We will conduct biological-behavioral surveillance in high-risk populations, with known bat contact, in community and

clinical settings to 1) identify risk factors for serological and PCR evidence of bat SARSr-CoVs; & 2) assess possible health effects of SARSr-CoVs infection in people. In addition, we will analyze bat-CoV serology against human-wildlife contact and exposure data to quantify risk factors and health impacts of SARSr-CoV spillover.

3. In vitro and in vivo characterization of SARSr-CoV spillover risk, we combined spatial and Phylogenetic analyses to identify the regions and viruses of public health concern. We will use S protein sequence data, infectious clone technology, in vitro and in vivo infection experiments, and receptor binding analysis to test the hypothesis that % divergence thresholds in S protein sequences predict spillover potential. We will combine these data with bat host distribution, viral diversity and phylogeny, the human survey of risk behaviors and illness, and serology to identify SARSr-CoV spillover risk hotspots across southern China. Together these data and analyses will be critical for the future development of public health interventions and enhanced surveillance to prevent the re-emergence of SARS or the emergence of a novel SARSr-CoV.

The project was launched in 2014 (1-June-2014) under Peter Daszak, the president of EcoHealth Alliance, New York, USA. The funding was provided by the National Institute of Allergy and Infectious Diseases. The amount of the financing was **$666,442.** Dr. Anthony Fauci is the NIAID director and Chief Medical Advisor to the President of the United States. The project was so finely designed with the precise end date of 30- June- 2026. The budget start date was July 24, 2019, and the budget end date was 30-June-2022

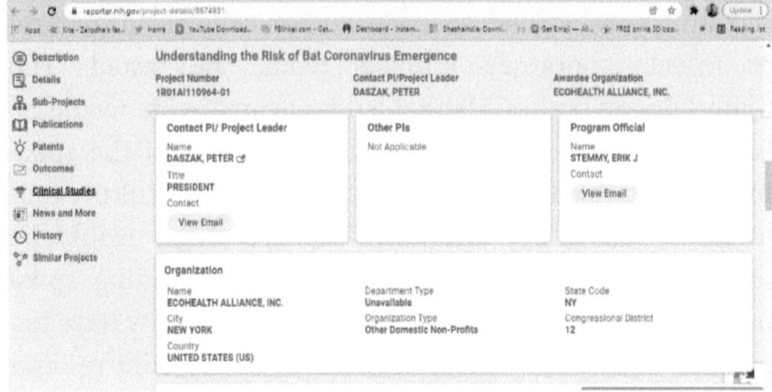

Description
Details
Sub-Projects
Publications
Patents
Outcomes
Clinical Studies
News and More
History
Similar Projects

Understanding the Risk of Bat Coronavirus Emergence

Project Number 1R01AI110964-01
Contact PI/Project Leader DASZAK, PETER
Awardee Organization ECOHEALTH ALLIANCE, INC.

Contact PI/ Project Leader
Name DASZAK, PETER
Title PRESIDENT
Contact View Email

Other PIs
Not Applicable

Program Official
Name STEMMY, ERIK J
Contact View Email

Organization
Name ECOHEALTH ALLIANCE, INC.
City NEW YORK
Country UNITED STATES (US)
Department Type Unavailable
Organization Type Other Domestic Non-Profits
State Code NY
Congressional District 12

NIH Report Screen Shot

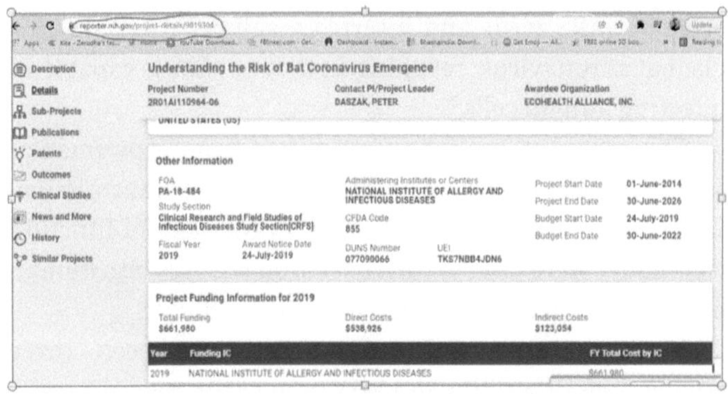

Description
Details
Sub-Projects
Publications
Patents
Outcomes
Clinical Studies
News and More
History
Similar Projects

Understanding the Risk of Bat Coronavirus Emergence

Project Number 2R01AI110964-06
Contact PI/Project Leader DASZAK, PETER
Awardee Organization ECOHEALTH ALLIANCE, INC.

Other Information
FOA PA-18-484
Study Section Clinical Research and Field Studies of Infectious Diseases Study Section[CRFS]
Fiscal Year 2019
Award Notice Date 24-July-2019
Administering Institutes or Centers NATIONAL INSTITUTE OF ALLERGY AND INFECTIOUS DISEASES
CFDA Code 855
DUNS Number 077090066
UEI TKS7NBB4JDN6
Project Start Date 01-June-2014
Project End Date 30-June-2026
Budget Start Date 24-July-2019
Budget End Date 30-June-2022

Project Funding Information for 2019
Total Funding $661,980
Direct Costs $538,926
Indirect Costs $123,054

Year	Funding IC	FY Total Cost by IC
2019	NATIONAL INSTITUTE OF ALLERGY AND INFECTIOUS DISEASES	$661,980

NIH Report-2

Translated into something approaching lay language, **Aim 3** states that de novo synthesis is used to construct a series of novel chimeric viruses, comprising recombinant hybrids using different spike proteins from each of a series of unpublished natural coronaviruses in an otherwise-constant genome of a bat coronavirus. The ability of the resulting novel viruses to infect human cells in culture and to infect laboratory animals would be tested. The underlying hypothesis is that a direct correlation would be found between the receptor-binding affinity of the spike protein and the ability to infect human cells in culture and to infect laboratory animals. This hypothesis would be tested by asking whether novel viruses encoding spike proteins with the highest receptor-binding affinity have the highest ability to infect human cells in culture and laboratory animals.

The WIV began its Gain of Function research program for bat coronaviruses in 2015. Using a natural virus, institute researchers made "substitutions in its RNA coding to make it more transmissible. They took a piece of the original SARS virus and inserted a snippet from a SARS-like bat coronavirus, resulting in a virus that is capable of infecting human cells."

This meant it could be transmitted from experimental animals to experimental animals by aerosol transmission, which could do humans. In other words, gain-of-function techniques were used to turn bat coronaviruses into human pathogens capable of causing a global pandemic.

In 2015, 2016, and 2017, there have been three publications describing the WIV gain of function research. The WIV, having learned both basic and traceless infectious-clone technology from joint research with a laboratory at the University of North Carolina (UNC) in 2015, initiated the construction of novel chimeric coronaviruses without UNC immediately after that. WIV's first publication on basic infectious-clone technology to construct novel chimeric coronaviruses at WIV appeared in 2016. WIV's first publication on the use of traceless, signature-free infectious-clone technology also appeared in 2016.

The Chinese government has proudly stated that the WIV "preserves more than 1,500 strains of the virus," the most extensive collection of bat and other coronaviruses in Asia. In Scientific American, the 2019 interview with Shi reports that the WIV had at least hundreds of individual strains. Chinese government authorities have said these numbers, which are being taken at face value here.

From 2004 on, the WIV published many dozens of partial or complete genome sequences of coronaviruses in their collection. For example, on June 1, Daszak and Shi published partial genetic sequences of 781 Chinese bat coronaviruses, more than one-third of which had never been published previously. There are also multiple published records of animal infection research with bat coronaviruses at the WIV. The WIV laboratory needs to use live viruses, not just RNA fragments, to carry out the research program described above. This contradicts two assertions made by some commentators that Shi worked only with RNA fragments and that her laboratory did not maintain live viruses. On May 24, 2020, the director of the WIV acknowledged that the laboratory did have "three live strains of bat coronaviruses on-site" but implied only three. Knowledgeable virologists assume that the number must be much higher, probably hundreds of live viral isolates.

November 11, 2021

"Darpan, there are now sufficient research papers available indicating that the Gain-of-function research was going on at the Wuhan Institute of Virology. But what makes SARS-CoV-2 lethal? One more thing is to be considered that if animals would start getting infected just like humans with it, I cannot imagine what would be the level of its catastrophic effect?" I asked Darpan

"I also asked the same question months back on a call to one of my microbiology professors in University!" Darpan replied to me quickly.

"So, what was his opinion?" I asked in eagerness.

He said to me, "To date, there are very few cases of such, but as the genetic structure of the virus changes and new variants emerges, anything could happen. "

"This would probably be the first of its kind pandemic that will spread from humans to animals!" I said to Darpan.

"Anyways, Darpan, but what is unique in this virus that makes it highly transmittable in humans and lethal too. And as we know that the death rate in India and across the world is more in the second wave than the first wave in 2020." I said to Darpan.

"One more thing to notice is China hasn't faced a second wave!" Darpan said.

"Interesting, isn't it? I do not say that the innocent Chinese population should have gone through the second wave, but surprisingly, China played well! That could be possible with early mass production of vaccine which they had already started by a company named CanSino earlier in ***November-2019*** ending." I told him.

"It could be a coincidence, couldn't it?" He replied.

"Yes, it could be a co Incidence. But yet I have another dot to connect with the dot of ***November-2019.*** A wall street journal article published on May 23, 2021,*By* Michael R. Gordon, Warren P. Strobel, *and* Drew Hinshaw that three of the WIV researchers were fell sick in ***November-2019*** with Covid-19 like symptoms. That I will share in my book." I replied with confidence to him.

Three researchers from China's Wuhan Institute of Virology became sick enough in November 2019 that they sought hospital care, according to a previously undisclosed U.S. intelligence report that could add weight to growing calls for a fuller probe of whether the Covid-19 virus may have escaped from the laboratory. The details of the reporting go beyond a State Department fact sheet, issued during the final days of the Trump administration, which said that several researchers at the lab, a center for the study of coronaviruses and other pathogens, became sick in autumn 2019 "with symptoms consistent with both Covid-19 and common seasonal illness."

The disclosure of the number of researchers, the timing of their illnesses, and their hospital visits come on the eve of a meeting of the World Health Organization's decision-making body, which is expected to discuss the next phase of an investigation into Covid-19's origins.

Current and former officials familiar with the intelligence about the lab researchers expressed differing views about the strength of the supporting evidence for the assessment. One person said that it was provided by an international partner and was potentially significant but still needed further investigation and additional corroboration.

The State Department fact sheet issued during the Trump administration, which drew on classified intelligence, said that the "U.S. government has reason to believe that several researchers inside the WIV became sick in autumn 2019, before the first identified case of the outbreak, with symptoms consistent with both Covid-19 and seasonal illnesses."

The Warning

In January 2018, the U.S. Embassy in Beijing took the unusual step of repeatedly sending U.S. science diplomats to the Wuhan Institute of Virology (WIV), which had in 2015 become China's first laboratory to achieve the highest level of international bioresearch safety (known as BSL-4). WIV issued a news release in English about the last of these visits, which occurred on March 27, 2018. The U.S. delegation was led by Jamison Fouss, the consul general in Wuhan, and Rick Switzer, the embassy's environment, science, technology, and health counselor. Last week, WIV erased that statement from its website, though it remains archived on the Internet.

The U.S. officials learned during their visits that concerned them so much then they dispatched two diplomatic cables categorized as Sensitive but Unclassified back to Washington. The cables warned about safety and management weaknesses at the WIV lab and proposed more attention and help. The first cable, which I obtained, also warns that the lab's work on bat coronaviruses and their potential human transmission represented a risk of a new SARS-like pandemic.

"During interactions with scientists at the WIV laboratory, they noted the new lab has a serious shortage of appropriately trained technicians, and investigators needed to operate this high-containment laboratory safely," states January 19, 2018, cable, which was drafted by two officials from the embassy's environment, science and health sections who met with the WIV scientists.

As the cable noted, the U.S. visitors met with Shi Zhengli, the head of the research project, who had been publishing studies related to bat coronaviruses for many years. In

November 2017, just before the U.S. officials' visit, Shi's team had published research showing that horseshoe bats they had collected from a cave in Yunnan province were very likely from the same bat population that spawned the SARS coronavirus in 2003.

Sources familiar with the cables said they were meant to sound an alarm about the grave safety concerns at the WIV lab, especially regarding its work with bat coronaviruses. The embassy officials were calling for more U.S. attention to this lab and more support to help it fix its problems.

"The cable was a warning shot," one U.S. official said. "They were begging people to pay attention to what was going on."

In the next chapter, we will learn more about the COVID-19 virus. Then, we will try to understand what makes Covid-19 more transmittable and lethal among the SARS- CoV family and why SARS-CoV-2 has been believed to be more likely laboratory engineered rather than naturally occurring.

VII
The Killer Couple

"Condemnation without investigation is the height of ignorance."

-Albert Einstein

"An investigator should have a robust faith- and yet not believe."

-Claude Bernard

12th November 2021

"I had done some research after our conversation last night, and I found that the SARS-CoV-2 has a unique Furin cleavage site. Many researchers around the world are speculating that this furin cleavage site is possibly deliberately engineered to increase transmissibility and infectivity among humans", Darpan said to me.

"The things you are saying are very scientific to me. Can you translate it to a layman's language?" I asked him.

"Sure!" He replied.

"Let's consider an example of a group of terrorists." Darpan continued further.

"Hey! if I am not wrong, we are discussing Coronavirus!" I interrupted him.

"Right, Nikunj, but it will be easier for a common man to understand how coronavirus enters and spreads inside a human body!" He replied.

"So, it is like a terrorist group with two guides. The two guides are the person who knows the exact location to invade. The others in that group have only one objective to spread terror after the invasion. The job of one of those two guides is to keep an eye around the location. The second guide will break the security of a particular location with an insider's help and allow the group of terrorists to enter that location. After the invasion, the terrorists will continue their preallocated job, and those two guides will leave that location." Darpan completed the story.

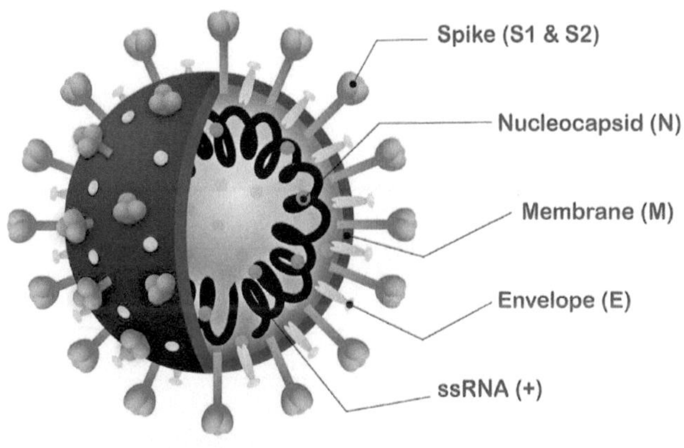

Corona Virus Structure

Now, look at this image,

Now, the above image of SARS-CoV-2 is a group of terrorists in which,

The spike proteins S1 and S2 are like those two guides. S1 is Receptor-Binding Domain (RBD), and S2 is Receptor-Binding Motif. S1 ensures binding to the host cell via angiotensin-converting enzyme 2 (ACE-2), and S2 ensures fusion and entry of virion into the host cell.

The Nucleocapsid (N) is like our core terrorist group, which will invade inside and spread terror. This Nucleocapsid will enter to host cell and will start multiplying the coronavirus.

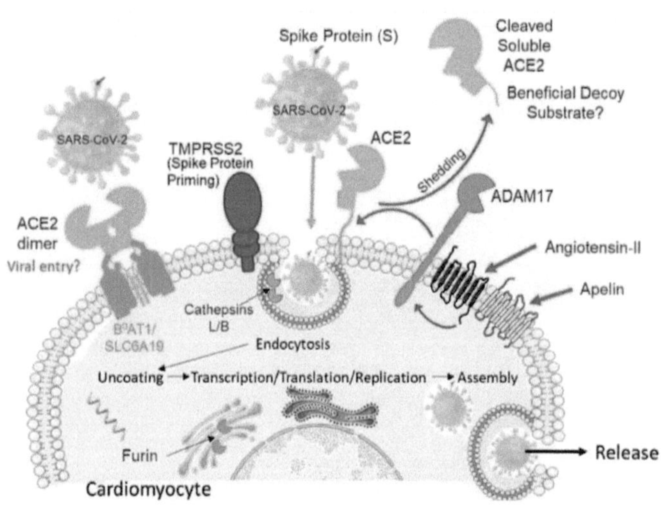

Image Credits: https://www.news-medical.net/news/20200709/SARS-CoV-2-a-dagger-to-the-aging-heart.aspx

Schematic diagram of the key proteins predicted from RNASeq data to be expressed by human cardiomyocytes. We propose SARS-CoV-2 binds initially to ACE2 (with the ACE2/B0AT1 complex as a potential second entry site).

TMPRSS2 priming of the spike protein S1 together with further protease activation by Cathepsins B and L facilitates viral cell entry and internalization by endocytosis. Furin may also have a role in this process. Internalization of the virus with ACE2 inhibits ACE2 carboxypeptidase activity that normally hydrolyzes Ang-II, [Pyr1]-apelin-13, and des-Arg9-bradykinin. ADAM17, present on the cell surface, cleaves ACE2 to the soluble form that circulates in the plasma and could act as a decoy substrate for the virus. ADAM17 may be regulated by Ang-II and apelin acting via their respective G-protein coupled receptors.

To infect host cells, the Spike protein on SARS-CoV-2 must bind ACE2 and undergo subsequent cleavage by TMPRSS2. This protease primes the Spike S1 subunit for the internalization of the virus. A second site on the S2 subunit is also cleaved by the enzyme furin. Once inside the cell, the virus undergoes endosomal processing by the cysteine proteases cathepsin L(CTSL) and cathepsin B (CTSB).

Yuri Deigin at medium.com explains the process of infection in layman's language. The part of that article under the title 'A Killer intro' is specially written for the spike protein and furin cleavage. It is impossible to ignore the introduction of a PRRA insert between S1 and S2: it sticks out like a splinter. This insert creates the furin cleavage site.

The protein consists of two parts, S1 and S2, of which S1 is responsible for primary contact with the receptor (recall Receptor Binding Domain / Motif), and S2 is responsible for fusion with the cell membrane and penetration into the cell. The fusion process is started by the fusion peptide marked in yellow, but to engage in its dirty deed, someone must cut the S protein at one of the sites marked by diamonds in the diagram above. Unfortunately, the virus does not have its own such "cutters," so it relies on various proteases of its victims. There are several types of such proteases, as can be deduced from the abundance of colors of those diamonds. But not all proteases are equal, and not all types of cells have proteases needed by the virus.

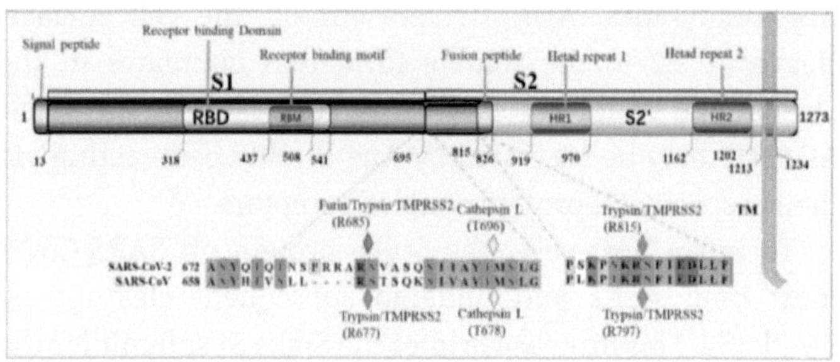

Image Source: https://yurideigin.medium.com/lab-made-cov2-genealogy-through-the-lens-of-gain-of-function-research-f96dd7413748

A coronavirus (CoV) infects the target cell by either cytoplasmic or endosomal membrane fusion. No matter what path it chooses, the final viral entry step involves the release of RNA into the cytoplasm for replication. Therefore, the fusion capacity of the CoV-S is a leading indicator of the infectivity of the corresponding virus. Consisting of S1 receptor-binding subunit and S2 fusion subunit, CoV-S needs to be primed through cleavage at S1/S2 site and S2′ site to mediate the membrane fusion (Fig. 1a). Previous studies have shown that an insertion of FCS consisting of multiple basic amino acids in the cleavage site of the haemagglutinin (HA) is associated with the high virulence of influenza viruses. Coincidentally, phylogenetic analysis of SARS-CoV-2 identified an insertion of RRAR

(FCS) at the S1/S2 site of SARS-CoV-2-S, which is absent in SARS-CoV and other SARS-related coronaviruses (SARSr-CoVs), particularly RaTG13, which has 96% identity of its genomic sequence to that of SARS-CoV-2 (Fig. 1b). Therefore, it has been speculated that this unique FCS may provide a gain-of-function, making SARS-CoV-2 easily enter the host cell for infection, thus efficiently spreading throughout the human population, compared to other lineage B betacoronaviruses.

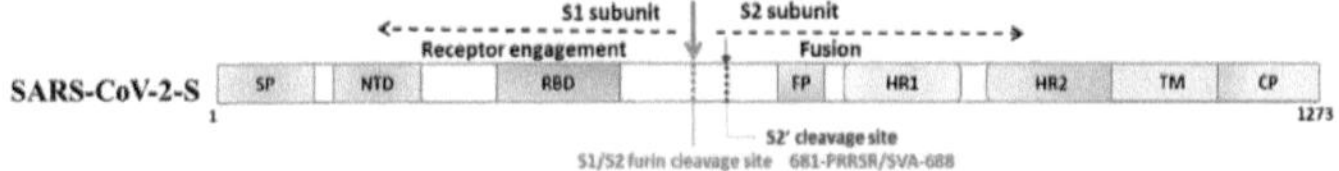

Figure a

	S1 | S2			Score
SARS-CoV	S1	YHTV --S--LLRST SQKSIVA	S2	< 0.5
RaTG13	S1	YQTQ TNS----RSV ASQSIIA	S2	< 0.5
SARS-CoV-2	S1	YQTQ TNSPRRARSV ASQSIIA	S2	0.62
SARS-CoV-2-m1	S1	YQTQ TNSPSSARSV ASQSIIA	S2	< 0.5
SARS-CoV-2-m2	S1	YQTQ TNS----RSV ASQSIIA	S2	< 0.5
SARS-CoV-2-m3	S1	YHTV --S--LLRST SQKSIVA	S2	< 0.5
SARS-CoV-m1	S1	YQTQ TNSPRRARSV ASQSIIA	S2	0.62
SARS-CoV-m2	S1	YQTQ TNSPSSARSV ASQSIIA	S2	< 0.5

Figure b

a. Schematic representation of SARS-CoV-2 S protein and the location of S1/S2 and S2′ cleavage site. SP, signal peptide; FP, fusion peptide; HR, heptad repeat domain; TM, transmembrane domain; CP, cytoplasmic domain. **b.** Mutated SARS-CoV-2 S proteins with a mutation in S1/S2 region, including SARS-CoV-2-m1 (mutating "RRAR" into "SSAR"), SARS-CoV-2-m2 (deleting four amino acids, "PRRA"), SARS-CoV-2-m3 (replacing "QTQTNSPRRARSVASQSII" in SARS-CoV-2 with "HTVSLLRSTSQKSIV" derived from SARS-CoV), SARS-CoV-m1 (replacing "HTVSLLRSTSQKSIV" in SARS-CoV with "QTQTNSPRRARSVASQSII" derived from SARS-CoV-2), and SARS-CoV-m2 (mutating "RRAR" in SARS-CoV-m1 into "SSAR"). Prediction scores for the S1/S2 furin cleavage site in S protein were analyzed by using the ProP 1.0 server

Later, a group of scientists led by Bryan A. Johnson experimented with a mutant of SARS-CoV-2 lacking Furin cleavage site in its spike protein. This mutant virus replicated with faster kinetics and improved fitness in Vero E6 cells. The mutant virus also had reduced spike protein processing compared to wild-type SARS-CoV-2. In contrast, the ΔPRRA had reduced replication in Calu3 cells, a human respiratory cell line, and ten had attenuated disease in a hamster pathogenesis model. Despite the reduced disease, the ΔPRRA mutant offered robust protection from SARS-CoV-2 rechallenge. Importantly, plaque reduction neutralization tests (PRNT) with COVID-19 patient sera and monoclonal antibodies against the receptor-binding domain found a shift, with the mutant virus resulting in consistently reduced PRNT titers. Together, these results demonstrate a critical role in inserting the furin cleavage site in SARS-CoV-2 replication and pathogenesis. In addition, these findings illustrate the importance of this insertion in evaluating neutralization and other downstream SARS-CoV-2 assays.

Much attention has been drawn to the origin of SARS-CoV-2, the causative agent of COVID-19. One notable feature of SARS-CoV-2 is a four-amino acid insert starting with proline (SPRRAR|S) at the junction of the receptor-binding (S1) and fusion (S2) domains of the spike protein. Following the release of the SARS-CoV-2 genome, several groups identified this insert as a potential cleavage site for the protease furin—the insert has also been referred to as a polybasic site and proposed to be part of the proximal origin of the ongoing pandemic.

Proteolytic cleavage is widely used to activate the fusion machinery of viral glycoproteins. However, studies on coronaviruses—including SARS-CoV-2—have shown that activation of the spike proteins of these particular viruses is not straightforward and is often a complex process involving more than one cleavage event at specific sites with the involvement of several host proteases.

As such, any consideration of spike protein origin and function needs to be considered along with the natural history of coronaviruses and how the spike protein adapts to a milieu of different species, tissue types, and cell types.

For coronaviruses, furin cleavage sites at the interface of the S1 and S2 domain are not unusual, being found widely in betacoronaviruses in the embeco lineage (which are considered to be of rodent origin) as well as in avian-origin gammacoronaviruses and certain feline and canine alphacoronaviruses (with an unknown origin). Furin cleavage sites are also found in certain bat-origin MERS-like merbecovirises, but not—except for SARS-CoV-2—in the sarbecovirus lineage. The presence of a furin cleavage motif at the SARS-CoV-2 S1–S2 interface is highly unusual, leading to the smoking gun hypothesis of manipulation that has recently gained considerable attention as a possible origin SARS-CoV-2.

SARS-CoV-2 is the only Sarbecovirus to contain an FCS (Coutard et al., 2020). Indeed, no CoV with a spike protein sequence homology greater than 40% to SARS-CoV-2 has an FCS (Wu, C. et al., 2020). The presence of an arginine at the third position P3 before the FCS increases the efficiency of the FCS tenfold. Its presence is rare, occurring in only 5 out of 132 known FCSs. (Lemmin et al., 2020)

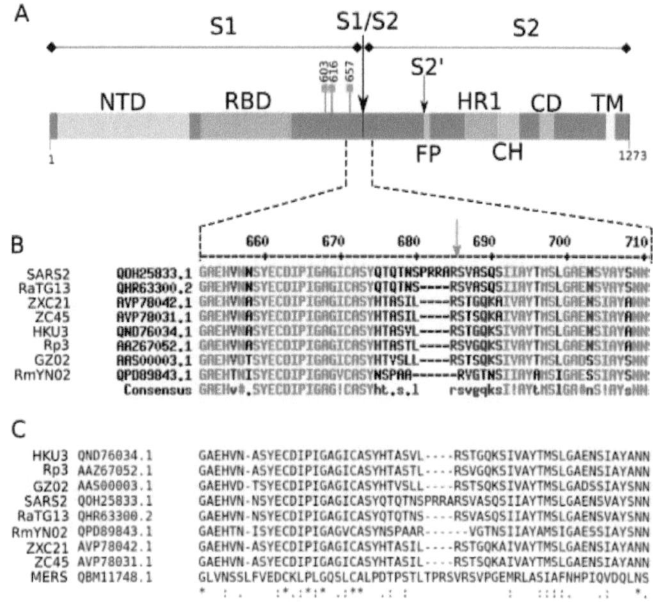

Fig. 1. Schematic representation of the SARS-CoV-2's spike protein showing subunits and domains as well as local sequence alignments with other Betacoronaviruses. A) Spike domains: N terminal domain (NTD); Receptor binding domain (RBD); Fusion peptide (FP); Hetapad repeat 1 (HR1); Central helix (CH); Connector domain (CD), Transmembrane Domain (TM). Arrows indicate the S1▾S2 and S2′ cleavage sites. Green boxes indicate the location of N-glycosylated residues proximal to FCS. B)

MultAlin alignments using SARS-CoV-2 spike protein sequence numbering reference (Corpet 1988). (http://www.sacs.ucsf.edu/cgi-bin/multalin.py). C) SACS ClustalW multiple sequence alignment (http://www.sacs.ucsf.edu/cgi-bin/clustalw.py). Definition and accession numbers as follows: SARS2: SARS-CoV-2 Wuhan-Hu-1 (QOH25833.1); RaTG13 (QHR63300.2); ZXC21: bat-SL-CoVZXC21 (AVP78042.1); ZC45: bat-SL-CoVZC45 (AVP78031.1); HKU3: Bat SARS coronavirus HKU3 (QND76034.1); Rp3: Rp3/2004 (AAZ67052.1); GZ02: SARS coronavirus GZ02 (AAS00003.1); RmYN02: Bat coronavirus RmYN02 (QPD89843.1); MERS: middle East respiratory syndrome-related coronavirus (QBM11748.1). SARS-CoV-2 referenced sequence indexes shown

Another unique feature of SARS-CoV-2 compared to related CoVs is a longer loop containing the S1/S2 cleavage site (Lemmin et al., 2020): it is at least 4 amino acids around the site containing the FCS than any other known Sarbecovirus. The combination of FCS and extended loop length facilitate SARS-CoV-2 activation by protease TMPRSS13 and TMPRSS2, albeit at one-third the effectiveness of TMPRSS2 (Laporte et al., 2020). Mutants with a deleted 'PRRA' insert or a shortened FCS loop with deleted preceding amino acids 'QTQTN' abrogated the effectiveness of both TMPRSS2 and TMPRSS13 for facilitating cleavage (Laporte et al., 2020).

Moreover, the FCS in SARS-CoV-2 is coded by rare codons, leaving it not in-frame with the rest of the sequence, thus violating the rules of the copy choice recombination mechanisms that postulate in-frame insertions. Additionally, its insertion causes a peculiar split of one of the codons, serine (TCA), compared with the close relatives MP789 and RaTG13 (Segreto & Deigin, 2020). The recent acquisition of the FCS by SARS-CoV-2 via a natural insert was proposed by Wu and Zhao (2020) based on the existence of FCS in other, not closely related Betacoronaviruses with different loop positions to SARS-CoV-2 and the existence of a partial natural insert in the same region in RmYN02 (Zhou, H. et al., 2020). However, the reliability of the conclusions of Zhou, H. et al. (2020) has been questioned by Deigin and Segreto (2020). They particularly challenge the claim that RmYN02 has an insertion around the site of the FCS insertion in SARS-CoV-2 and instead point to a two amino acid deletion in RmYN02 at that locus. Therefore, RmYN02 should not be used as evidence of the natural origin of SARS-CoV2's FCS until its claimed insertion is properly validated.

A minimal FCS could have evolved via a single point mutation T678R (Li, W. et al., 2015), which is evolutionarily more economical than a complete 12nt insertion of 'PRRA.' A multi basic cleavage site is also plausible with an additional mutation N679R. We note that deletions but not insertions frequently happen at the S1/S2 junction of SARS-CoV-2 during serial cell passage (Peacock et al., 2020) and have also been detected in strains isolated from hamsters and humans (Lau S-Y. et al. 2020; Liu Z. et al. 2020). The acquisition of the FCS via a natural insert, when an FCS could have evolved far more easily through point mutation, we believe is highly unlikely. Because the presence and coding sequence is important for pathogenesis, host range, and cell tropism (Nagai et al., 1993; Millet et al., 2015), adding an FCS into viruses has been an active area of gain-of-function research. An FCS can be easily inserted using seamless technology (Yount et al., 2002; Sirotkin and

Sirotkin 2020) without any need for cell passage, as previously performed in experiments on virulence and host tropism (Cheng et al. 2019). Insertions to change the properties of SARS-r CoV viruses are documented by Ren et al. (2008) and Wang et al. (2008). Natural mutations have a very low probability of resulting in a stretch of 12 amino acids coding for an optimized FCS without any known intermediate form in Sarbecovirus; an artificial insertion of the FCS in SARS-CoV-2 may provide a more parsimonious explanation for its presence than natural evolution.

SARS-CoV-2 differs from its closest relative RaTG13 by a few key characteristics. The most striking difference is the acquisition in the spike protein of SARS-CoV-2 of a cleavage site activated by host-cell enzyme furin, previously not identified in other beta-CoVs of lineage band similar to that of Middle East respiratory syndrome (MERS) coronavirus. Host protease processing plays a pivotal role as a species and tissue barrier, and engineering the cleavage sites of CoV spike proteins modify virus tropism and virulence.

The polybasic furin site in SARS-CoV-2 was created by a 12-nucleotide insert TCCTCGGCGGGC coding for a PRRA amino acid sequence at the S1/S2 junction (Figure). Interestingly, the two joint arginines are coded by two CGGCGG codons, rare for these viruses: only 5% of arginines are coded by CGG in SARS-CoV-2 or RaTG13, and CGGCGG in the new insert is the only doubled instance of this codon in SARS-CoV-2. In addition, the CGGCGG insert includes a *Fau*I restriction site, of which there are six instances in SARS-CoV-2 and four instances in RaTG13 (and two in MP789). The advantageous location of the *Fau*I site could allow using restriction fragment length polymorphism (RFLP) techniques for cloning or screening for mutations, as the new furin site is prone to deletions *in vitro*.

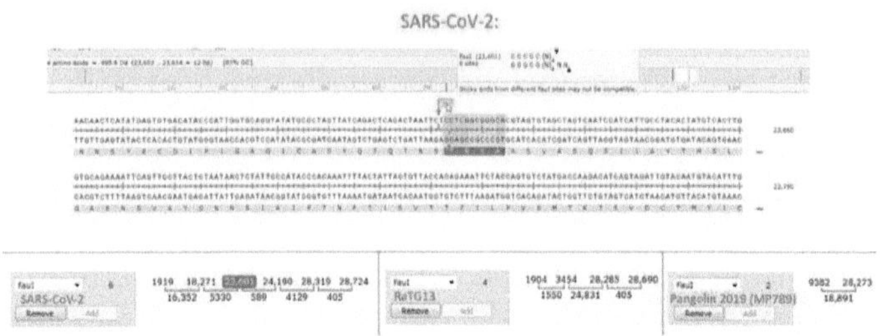

Therefore, SARS-CoV-2 remains unique among its beta CoV relatives due to a polybasic furin site at the S1/S2 junction and the four amino acid insert PRRA that had created it. The insertion causes a split in the original codon for serine (TCA) in MP789 or RaTG13 to give part of a new codon for serine (TCT) and part of the amino acid alanine (GCA) in SARS-CoV-2.

And we can not forget the highly specific aim of Peter Daszak and Shi Zhengli's team for the project of '*Understanding the risk of Bat Coronavirus emergence*' The aim clearly stated *that 'In vitro and in vivo characterization of SARSr-CoV spillover risk, coupled with spatial and phylogenetic analyses to identify the regions and viruses of public health concern. We will use S protein sequence data, infectious clone technology, in vitro and in vivo infection experiments, and analysis of receptor binding to test the hypothesis that % divergence thresholds in S protein sequences predict spillover potential.'*

VIII

The Invisible Weapon

"When there is not enough to eat, people starve to death. It is better to let half of the people die so that the other half can eat their fill." ~ Mao Zedong

"All political power comes from the barrel of a gun. The communist party must command all the guns; that way, no guns can ever be used to command the party." ~ Mao Zedong

"Deaths have benefits. They can fertilize the ground." ~ Mao Zedong

"Don't make a fuss about a world war. At most, people die... Half the population wiped out - this happened quite a few times in Chinese history... It's best if half the population is left, next best one-third." ~ Mao Zedong

'China Overtakes U.S. as world's richest Nation as global wealth surges' – India Today (Digital News platform) the most surprising headline on 16th November 2021.

Why surprising?

Ans. In the peak duration of the Covid-19 pandemic of December-2019 to 2021, most of the countries' economy was struggling and sinking, but in contrast, the Chinese economy became the world's largest economy.

When I completed the 7th chapter of this book, I was confused about what to write next. The buzz was around the Covid-19 that it could be possible that the Chinese army had developed SARS-CoV-2, and later it escaped from the laboratory via an incident.

An article published in Australian media with the headline, ***'Covid as biological war? China paper needs answers'*** according to that news article. According to that article, People's Liberation Army scientists and senior Chinese health officials set out their ideas "in a document that predicted a third world war would be fought with biological weapons." That document, written in 2015, was obtained by US State Department officials. It described SARS coronaviruses as heralding a "new era of genetic weapons." The viruses, it said, could be "artificially manipulated into an emerging human disease virus, then weaponized and unleashed in a way never seen before."

China has a history with biowarfare. In the 1930s, through the second China-Japan war and till the end of World War II, Unit 731 carried out horrific human experiments on the countless prisoners and villagers at Harbin, Northeastern China. The Unit was officially known as the Epidemic Prevention and Water Purification Department of the Kwantung Army.

As per reports, Japan's biological warfare program was started in the 1930s, after biological warfare was banned in the Geneva Convention in 1925. The Kwantung Army, which had at that time controlled large swathes of China, made the headquarters near the Pingfang district in Harbin and evicted around eight villages to make their facilities. One of the main reasons for selecting Harbin was that 'test subjects' were readily available. Unit 731 was led by General Shiro Ishii, a combat medical officer of formidable reputation.

Unit 731 members, image from 2017 NHK Documentary

13th November 2021

"Darpan, from 1930 to the end of world war 2 Chinese were the victim of bio-weapons. We know a well-documented history of the Jews' genocide using different poisonous gases and chemicals. But there are stories of atrocity that have gone largely undiscussed. One of the lesser-known tales is that of Unit 731." I started the conversation.

"Unit 731? What was that?" Darpan asked me curiously.

"Unit 731 was the name of the Imperial Japanese Army's covert biological and chemical warfare division. In the 1930s, through the second China-Japan war and till the end of World War II, Unit 731 carried out horrific human experiments on the countless prisoners and villagers at Harbin, Northeastern China. The Unit was officially known as the Epidemic Prevention and Water Purification Department of the Kwantung Army." I replied to him.

"So what is the relationship between bioweapons of China and Unit 731?" He asked.

"Japanese Unit 731 brutally tortured Chinese PLA prisoners and also tested 'Plague' like pathogens! They used to send infected people among healthy people to study infections. As a result, some war prisoners got infected with multiple pathogens to examine cross-reactions of different diseases." I replied.

"Horrible!" he replied.

I continued the topic, "The height of brutality was when doctors under Shiro Ishii, to study the spread of sexually transmitted diseases, Unit 731 doctors forced subjects to rape and impregnate female prisoners. Some women were forcefully impregnated to study disease progression during pregnancy and fetal transmission. Foetus, newborns, were test subjects too."

"Right, but my friend, our topic is SARS-CoV-2. Do you want to say that the SARS-CoV-2 was also somewhere related to the Chinese Bioweapons program?" Darpan was curious to know.

"We will discuss that part too. Because my friend, all you need is evidence, and without evidence, everything seems like a theory. Also, it is our right to access and disclose such important evidence." I replied him, leaving curious at that moment.

"Ok, so let's start where we stopped." He said.

Unit 731 experimenting on prisoners, image via SCMP

"The most brutal tales of the atrocities of Unit 731 were the accounts of vivisection, cutting open live prisoners, often without anesthesia. In an NYT report, a former Unit 731 member has described how they cut open prisoners while they screamed under the condition of anonymity. The purpose of vivisection was to study the spread of pathogens into the internal organs of test subjects while they were alive." I said.

"What was the intention of Unit 731?" Darpan asked.

"The intention of Unit 731 was to develop 'Plague bombs,' using humans to spread diseases among japan's enemies. They experimented with plagues, anthrax, and cholera." I said, watching at my wristwatch.

"It's 2:00 AM," Darpan said to me, as he saw me when I was staring at the clock.

"So, are you feeling sleepy?" I asked him.

"Ok, let's close the lab. And go for a tea." I said.

Meanwhile, My wife called me. She asked me about when I was returning home. I told her that we would sit the whole night and discuss and research the SARS-CoV-2. I told her to sleep and not to wait for me. I put on my cell phone.

We left for Tea on our 'Darling,' my two-wheeler.

"Do you remember college days?" I asked Darpan.

"Any doubts? Those were the golden days. Late-night group studies, tea and snacks, class bunks, and whatnot. No one can forget those days." He replied with a big smile.

"Park here," I said.

"Wow, truly heaven," Darpan said.

"I will order," I said.

I ordered two *Elaichi* (Cardamom) flavorteas for us. The tea vendor told us it would take about 15 minutes to prepare. Meanwhile, we continued our discussion.

"So, where were we?" I asked.

"We were at vivisection." He answered.

"Oh yes, but it does not end here. Plague outbreaks were reported in Ningbo and Changde at that time. It was later found out that the outbreaks were Unit 731 experiments, where they had released disease vectors into populations." I described more about Unit 731.

"Interesting information, but what exactly do they want to achieve?" Darpan asked; meanwhile, our tea arrived.

"Wow, what fragrance! I could not stop myself now" I sipped the tea.

"The drink of heaven!" Darpan said.

"The Ultimate aim of Unit 731 was to start epidemics in the U.S.A. The bomb balloons were reportedly precursors of balloons that would have eventually carried plague bombs. When the USA dropped nuclear bombs on Hiroshima and Nagasaki, making it clear that the war was over, Unit 731 used dynamites to destroy its headquarters in China. It is believed that after that 'Biological Massacre' PLA started their own bioweapons program in 1949 when they established the People's Republic." I completed.

"So, we can not rule out that the Covid-19 could have produced for malevolent cause in a Lab. and accidentally released from it," Darpan said, completing his cup of tea.

"Not," I replied

I paid for the tea.

I asked the Tea vendor, "Did you recognize us? We are the same guys who used to come to have a cup of tea here. I think you are Shanti Bhai's son. Right?"

"You are right, sir," He replied.

"Where is Shanti Bhai?" Darpan and I asked him simultaneously.

"He is no more in this world!" he replied.

"Why what happened to him?" I asked rapidly.

"He passed away; he succumbed to the Coronavirus in May!" he replied with heaviness in his voice.

We expressed our compunction. We left the place and gave some extra money for moral support.

After Some days

Remember the article published in Australian media, ***'Covid as biological war? China paper needs answers'.*** The book obtained by the U.S. state department was 'The Unnatural Origin of SARS and New Species of Man-Made Viruses as Genetic Bio-weapons.' This book could have been proven as a final nail in the coffin of the Natural Emergence theory of SARS-CoV and almost for SARS-CoV-2 emergence.

But before reaching that conclusion, one must prove the Wuhan Institute of Virology and PLA (People's Liberation

Army).

In 2013, the American virologist Ralph Baric approached Zhengli Shi at a meeting. Shi had detected the genome of a new virus, called SHC014, one of the two closest relatives to the original SARS virus, but her team had not been able to culture it in the laboratory. Baric had developed a way around that problem—a "reverse genetics" technique in coronaviruses. Not only did it allow him to bring an actual virus to life from its genetic code, but he could mix and match parts of multiple viruses. He wanted to take the "spike" gene from SHC014 and move it into a genetic copy of the SARS virus he already had in his laboratory. The spike molecule lets a coronavirus open a cell and get inside it. The resulting chimera would demonstrate whether the spike of SHC014 would attach to human cells.

Baric asked Shi Zhengli if he could have the genetic data for SHC014. "She was gracious enough to send us those sequences almost immediately," he told media. So his team introduced the virus modified with that code into mice and a petri dish of human airway cells. Sure enough, the chimera exhibited "robust replication" in the human cells—evidence that nature was full of coronaviruses ready to leap directly to people.

After Two Years, on 9th November 2015, an article was published by a team of researchers led by Shi Zhengli and Ralph S Baric.

The Article clearly states,

'The emergence of severe acute respiratory syndrome coronavirus (SARS-CoV) and the Middle East respiratory syndrome (MERS)-CoV underscores the threat of cross-species transmission events leading to outbreaks in humans. Here we examine the disease potential of a SARS-like virus, SHC014-CoV, which is currently circulating in Chinese horseshoe bat

populations. Using the SARS-CoV reverse genetics system, ***we generated and characterized a chimeric virus expressing the spike of bat coronavirus SHC014 in a mouse-adapted SARS-CoV backbone****. The results indicate that group 2b viruses encoding the SHC014 spike in a wild-type backbone can efficiently use multiple orthologs of the SARS receptor human angiotensin-converting enzyme II (ACE2), replicate efficiently in primary human airway cells, and achieve in vitro titers equivalent to epidemic strains of SARS-CoV. Additionally, in vivo experiments demonstrate replication of the chimeric virus in mouse lungs with notable pathogenesis. Evaluation of available SARS-based immune-therapeutic and prophylactic modalities revealed poor efficacy; both monoclonal antibody and vaccine approach failed to neutralize and protect from infection with CoVs using the novel spike protein. Based on these findings, we synthetically re-derived an infectious full-length SHC014 recombinant virus and demonstrated robust viral replication in vitro and Vivo. Our work suggests a potential risk of SARS-CoV re-emergence from viruses currently circulating in bat populations.'*

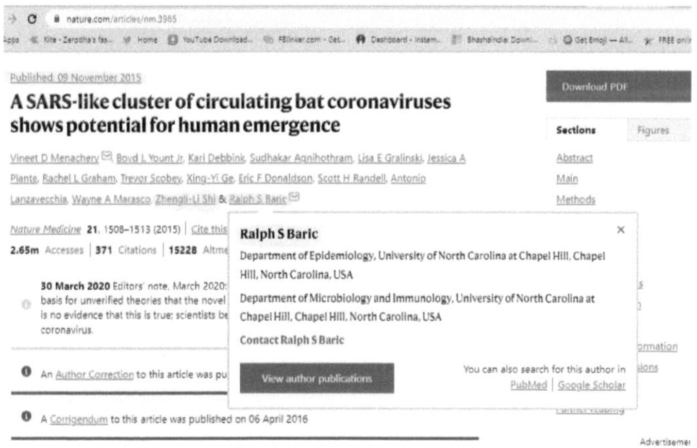
nature.com/articles/nm.3985

Published: 09 November 2015

Download PDF

A SARS-like cluster of circulating bat coronaviruses shows potential for human emergence

Sections | Figures

Vineet D Menachery, Boyd L Yount Jr, Kari Debbink, Sudhakar Agnihothram, Lisa E Gralinski, Jessica A Plante, Rachel L Graham, Trevor Scobey, Xing-Yi Ge, Eric F Donaldson, Scott H Randell, Antonio Lanzavecchia, Wayne A Marasco, Zhengli-Li Shi & Ralph S Baric

Abstract
Main
Methods

Nature Medicine **21**, 1508–1513 (2015) | Cite this

2.65m Accesses | **371** Citations | **15228** Altme

30 March 2020 Editors' note, March 2020: basis for unverified theories that the novel is no evidence that this is true; scientists be coronavirus.

Ralph S Baric

Department of Epidemiology, University of North Carolina at Chapel Hill, Chapel Hill, North Carolina, USA

Department of Microbiology and Immunology, University of North Carolina at Chapel Hill, Chapel Hill, North Carolina, USA

Contact Ralph S Baric

View author publications

You can also search for this author in PubMed | Google Scholar

An Author Correction to this article was pu

A Corrigendum to this article was published on 06 April 2016

Image Source Nature.com

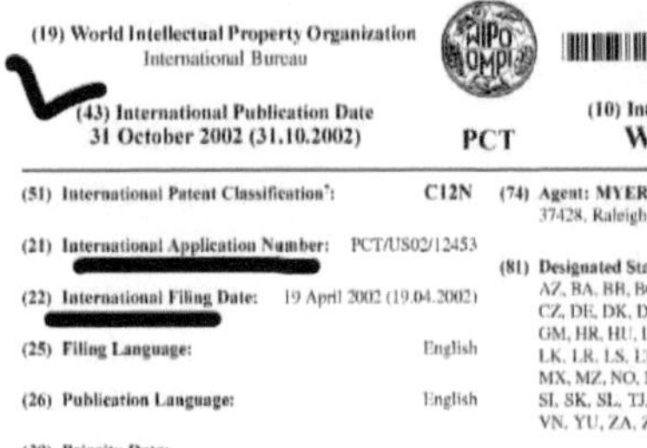

(12) INTERNATIONAL APPLICATION PUBLISHED UNDER THE PATENT COOPERATION TREATY (PCT)

(19) World Intellectual Property Organization
International Bureau

(43) International Publication Date
31 October 2002 (31.10.2002)

PCT

(10) International Publication Number
WO 02/086068 A2

(51) International Patent Classification[7]: C12N

(21) International Application Number: PCT/US02/12453

(22) International Filing Date: 19 April 2002 (19.04.2002)

(25) Filing Language: English

(26) Publication Language: English

(30) Priority Data:

60/285,320	20 April 2001 (20.04.2001)	US
60/285,318	20 April 2001 (20.04.2001)	US

(71) Applicant *(for all designated States except US)*: **THE UNIVERSITY OF NORTH CAROLINA AT CHAPEL HILL** [US/US]; 308 Bynum Hall, Campus Box 4105, Chapel Hill, NC 27599-4105 (US).

(72) Inventors: and
(75) Inventors/Applicants *(for US only)*: **CURTIS, Kristopher, M.** [US/US]; 417 Landerwood Lane, Chapel Hill NC 27514 (US). **YOUNT, Boyd** [US[illegible]; [illegible] Road, Chapel Hill, NC 27516 ([illegible]). **BARIC, Ralph, S.** [US/US]; 2600 Northstream Cour[illegible] [illegible]aw River, NC 27258-9529 (US).

(74) Agent: **MYERS BIGEL SIBLEY & SAJOVEC**; PO Box 37428, Raleigh, NC 27627 (US).

(81) Designated States *(national)*: AE, AG, AL, AM, AT, AU, AZ, BA, BB, BG, BR, BY, BZ, CA, CH, CN, CO, CR, CU, CZ, DE, DK, DM, DZ, EC, EE, ES, FI, GB, GD, GE, GH, GM, HR, HU, ID, IL, IN, IS, JP, KE, KG, KP, KR, KZ, LC, LK, LR, LS, LT, LU, LV, MA, MD, MG, MK, MN, MW, MX, MZ, NO, NZ, OM, PH, PL, PT, RO, RU, SD, SE, SG, SI, SK, SL, TJ, TM, TN, TR, TT, TZ, UA, UG, US, UZ, VN, YU, ZA, ZM, ZW.

(84) Designated States *(regional)*: ARIPO patent (GH, GM, KE, LS, MW, MZ, SD, SL, SZ, TZ, UG, ZM, ZW), Eurasian patent (AM, AZ, BY, KG, KZ, MD, RU, TJ, TM), European patent (AT, BE, CH, CY, DE, DK, ES, FI, FR, GB, GR, IE, IT, LU, MC, NL, PT, SE, TR), OAPI patent (BF, BJ, CF, CG, CI, CM, GA, GN, GQ, GW, ML, MR, NE, SN, TD, TG).

Published:
— *without international search report and to be republished upon receipt of that report*

[illegible]r two-letter codes and other abbreviations, refer to the "Guid[illegible]nce Notes on Codes and Abbreviations" appearing at the beginning of each regular issue of the PCT Gazette.

Ralph Baric Patent Registeration

60/285,318 20 April 2001 (20.04.2001) US

(71) **Applicant** *(for all designated States except US)*: **THE UNIVERSITY OF NORTH CAROLINA AT CHAPEL HILL** [US/US]; 308 Bynum Hall, Campus Box 4105, Chapel Hill, NC 27599-4105 (US).

(72) **Inventors; and**
(75) **Inventors/Applicants** *(for US only)*: **CURTIS, Kristopher, M.** [US/US]; 417 Landerwood Lane, Chapel Hill, NC 27514 (US). **YOUNT, Boyd** [US/US]; 1625 Smith Level Road, Chapel Hill, NC 27516 (US). **BARIC, Ralph, S.** [US/US]; 2600 Northstream Court, Haw River, NC 27258-9529 (US).

KE, LS, MW, MZ, SD, SL, SZ, TZ, UG, ZM, ZW), Eurasian patent (AM, AZ, BY, KG, KZ, MD, RU, TJ, TM), European patent (AT, BE, CH, CY, DE, DK, ES, FI, FR, GB, GR, IE, IT, LU, MC, NL, PT, SE, TR), OAPI patent (BF, BJ, CF, CG, CI, CM, GA, GN, GQ, GW, ML, MR, NE, SN, TD, TG).

Published:
— *without international search report and to be republished upon receipt of that report*

For two-letter codes and other abbreviations, refer to the "Guidance Notes on Codes and Abbreviations" appearing at the beginning of each regular issue of the PCT Gazette.

WO 02/086068 A2

(54) **Title:** METHODS FOR PRODUCING RECOMBINANT CORONAVIRUS

(57) **Abstract:** A helper cell for producing an infectious, replication defective, coronavirus (or more generally nidovirus) particle cell comprises (a) a nidovirus permissive cell; (b) a nidovirus replicon RNA comprising the nidovirus packaging signal and a heterologous RNA sequence, wherein the replicon RNA further lacks a sequence encoding at least one nidovirus structural protein; and (c) at least one separate helper RNA encoding the at least one structural protein absent from the replicon RNA, the helper RNA(s) lacking the nidovirus packaging signal. The combined expression of the replicon RNA and the helper RNA in the nidovirus permissive cell produces an assembled nidovirus particle which comprises the heterologous RNA sequence, is able to infect a cell, and is unable to complete viral replication in the absence of the helper RNA due to the absence of the structural protein coding sequence in the packaged replicon. Compositions for use in making such helper cells, along with viral particles produced from such cells, compositions of such viral particles, and methods of making and using such viral particles, are also disclosed.

Patent Registeration Document, Source: Anonymous

"A Picture is Worth Ten Thousand words" – Chinese Proverb

"So what is next? We were also finding the links between PLA and WIV." Darpan asked me.

"I know that will also be established," I said.

From 17th – 19th May, China-US jointly organized a workshop on the challenges of emerging infections. In that workshop, U.S. – China scientists participated but Pakistani, and Kenya's experts also participated. All is good but Pakistani experts? The whole world knows that Pakistani nuclear scientist Dr. A.Q. Khan theft Uranium Enrichment Technology and sold it to nations like North Korea and, god knows how, many other countries. But this could devastate more and that too in a very silent way.

"Though there is no linkage between PLA and WIV emerging!" Darpan said to me when he called me on next day.

"That is right, my friend, but I am still connecting some dots and will revert you soon!" I replied him.

Group Photo Including Ralph Baric, Shi Zheng Li Peiyon Shi

Ralph S Baric with other Chinese Virologist at Wuhan who worked on Covid 19 virus

Key figures in CCP virus research and development are indispensable:

Military Medical Department Zhou Yusen and Zhao Guangyu, Hong Kong's mysterious figure Zheng Bojian. The master of this technology is Baric, who drained this technology to Wuhan Institute of virology and the Military Medical Department through the New York Blood Center (Du Lanying, Jiang Shibo, and Li Fang). Eventually, the Zhoushan virus was weaponized by military-civilian integration.

Image Source An anonymous E-mail

Image Source An anonymous E-mail

Cleavage of spike protein of SARS coronavirus by protease factor Xa is associated with viral infectivity

…ng Du [a], Richard Y. Kao [a], Yusen Zhou [b], Yuxian He [c], Guangyu Zhao [a, b], Charlotte Wong [a], Shibo Jiang [c], Kwok-…g Yuen [a], Dong-Yan Jin [d], Bo-Jian Zheng [a]

Show more

+ Add to Mendeley Share Cite

https://doi.org/10.1016/j.bbrc.2007.05.092 Get rights and content

From an Anonyous source

Nature Public Health Emergency Collection

Nat Rev Microbiol. 2009; 7(3): 226–236.
Published online 2009 Feb 9. doi: 10.1038/nrmicro2090

PMCID: PMC2750777
NIHMSID: NIHMS132953
PMID: 19198616

The spike protein of SARS-CoV — a target for vaccine and therapeutic development

Lanying Du,[1] Yuxian He,[1] Yusen Zhou,[2] Shuwen Liu,[3] Bo-Jian Zheng,[4] and Shibo Jiang[1]

‣ Author information ‣ Copyright and License information Disclaimer

This article has been cited by other articles in PMC.

Abstract

Severe acute respiratory syndrome (SARS) is a newly emerging infectious disease caused by a novel coronavirus, SARS-coronavirus (SARS-CoV). The SARS-CoV spike (S) protein is composed of two subunits; the S1 subunit contains a receptor-binding domain that engages with the host cell receptor angiotensin-converting enzyme 2 and the S2 subunit mediates fusion between the viral and host cell membranes. The S protein plays key parts in the induction of neutralizing-antibody and T-cell responses, as well as protective immunity, during infection with SARS-CoV. In this Review, we highlight recent advances in the development of vaccines and therapeutics based on the S protein.

Image Source: Nature

Through the key clue of "restriction site," a complete chain of evidence has been formed through literature search.

Yusen Zhou, Bo-Jiang Zheng, and the same team in the above picture were working particularly spike proteins of SARS-CoV in 2009. And a research article related to this was published on PubMed on 9th February 2009.

Also, Yusen's article on 'The spike protein of SARS-CoV — a target for vaccine and therapeutic development' The last paragraph is very interesting; it states that,

'Early clinical studies based on such strategies have been carried out, but it is difficult to push the clinical trials of these candidate vaccines and therapeutics forward due to a lack of SARS-CoV-infected subjects and insufficient financial support. Thus, most big pharmaceutical companies have no interest in developing SARS vaccines and therapeutics because of the concern of profitability. However, studies on SARS will provide important information for designing novel strategies for prophylaxis and therapies of other newly emerging infections caused by enveloped viruses with class I fusion proteins.'

The Yan Report

It describes the genomic, structural, medical, and literature evidence, which, when considered together, strongly contradicts the natural origin theory. The evidence shows that SARS-CoV-2 should be a laboratory product created using bat coronaviruses ZC45 and/or ZXC21 as a template and/or backbone. Building upon the evidence, we further postulate a synthetic route for SARS-CoV-2, demonstrating that the laboratory creation of this coronavirus is convenient and can be accomplished in approximately six months. Our work emphasizes the need for an independent investigation into the relevant research laboratories. It also argues for a critical look into certain recently published data, which, albeit problematic, was used to support and claim a natural origin of SARS-CoV-2.

The three Yan reports used scientific evidence and analyses to prove that SARS-CoV-2 is an

Unrestricted Bioweapon was created by military scientists of the Chinese Communist Party (CCP) regime. These reports have played a pivotal role in revealing the true identity of the ongoing *Unrestricted Biowarfare.*

In the Yan report, Under sub-section 2.2, the method for creating SARS-CoV-2 has been described in five steps.

Step -1: Engineering the RBM of the Spike for hACE2-binding (1.5 months)

Step 2: Engineering a furin-cleavage site at the S1/S2 junction (0.5 months)

Step 4: Produce the designed viral genome using reverse genetics and recover live viruses (0.5 months)

Step 5: Optimize the virus for fitness and improve its hACE2-binding affinity *in vivo* (2.5-3 months)

The notable thing in the above five steps is the duration of time for manufacturing SARS-CoV-2 its around six months only. We know that to evolve a virus naturally would require decades. IN A TV INTERVIEW, the WIV director Yang Yi Wang said that even RaTG13 the closest relative of SARS-CoV-2, would take about 50 years to evolve into SARS-CoV-2. (she used the comment of Edvard Holmes, a leading virologist) But according to the Yan report, the same RaTG13 could be converted into SARS-CoV-2 within months!

Also, in the final remarks of Yan report is,

Motives aside, the following facts about SARS-CoV-2 are well-supported:

1. If it was a laboratory product, the most critical element in its creation, the backbone/template virus

(ZC45/ZXC21), is owned by military research laboratories.

2. The genome sequence of SARS-CoV-2 has likely undergone genetic engineering, through which

the virus has gained the ability to target humans with enhanced virulence and infectivity.

PLA members and WIV scientists shared multiple research articles see below:

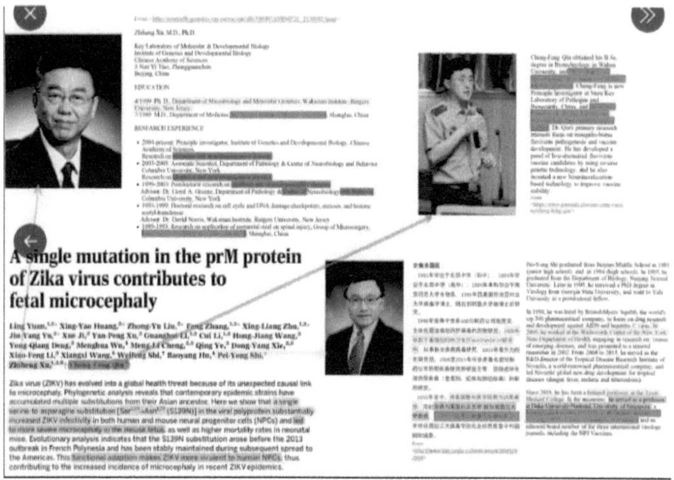

A single mutation in the prM protein of Zika virus contributes to fetal microcephaly

Ling Yuan, Xing-Yao Huang, Zhong-Yu Liu, Feng Zhang, Xing-Liang Zhu, Jiu-Yang Yu, Xue Ji, Yan-Peng Xu, Guanghui Li, Cui Li, Hong-Jiang Wang, Yong-Qiang Deng, Menghua Wu, Meng-Li Cheng, Qing Ye, Dong-Yang Xie, Xiao-Feng Li, Xiangxi Wang, Weifeng Shi, Baoyang Hu, Pei-Yong Shi, Zhiheng Xu, Cheng-Feng Qin

Zika virus (ZIKV) has evolved into a global health threat because of its unexpected causal link to microcephaly. Phylogenetic analysis reveals that contemporary epidemic strains have accumulated multiple substitutions from their Asian ancestor. Here we show that a single serine to asparagine substitution [$Ser^{139} \rightarrow Asn^{139}$ (S139N)] in the viral polyprotein substantially increased ZIKV infectivity in both human and mouse neural progenitor cells (NPCs) and led to more severe microcephaly in the mouse fetus, as well as higher mortality rates in neonatal mice. Evolutionary analysis indicates that the S139N substitution arose before the 2013 outbreak in French Polynesia and has been stably maintained during subsequent spread to the Americas. This functional adaption makes ZIKV more virulent to human NPCs, thus contributing to the increased incidence of microcephaly in recent ZIKV epidemics.

Image Source: An Anonymous E-mail

Pei-Yong Shi and Shi ZhengLi from WIV have close ties with Ralph S Baric. In earlier images shared in this chapter, Pei-Yong Shi and ZhengLi share the stage. Pei Yong-Shi and Cheng Feng Qin (PLA) [see above image] were involved in the research related to various viruses.

Cheng Feng Qin is from the Department of Virology, State Key Laboratory of Pathogen and Biosecurity, Beijing Institute of Microbiology and Epidemiology, 100071 Beijing, China.

Cheng Feng Qin and Pei-Yong Shi shared a patent registration of the Novel attenuated Dengue virus strain for vaccine application in August 2015. In that patent registration, Cheng Feng Qin showed a fake affiliation of Singapore! See Below Image.

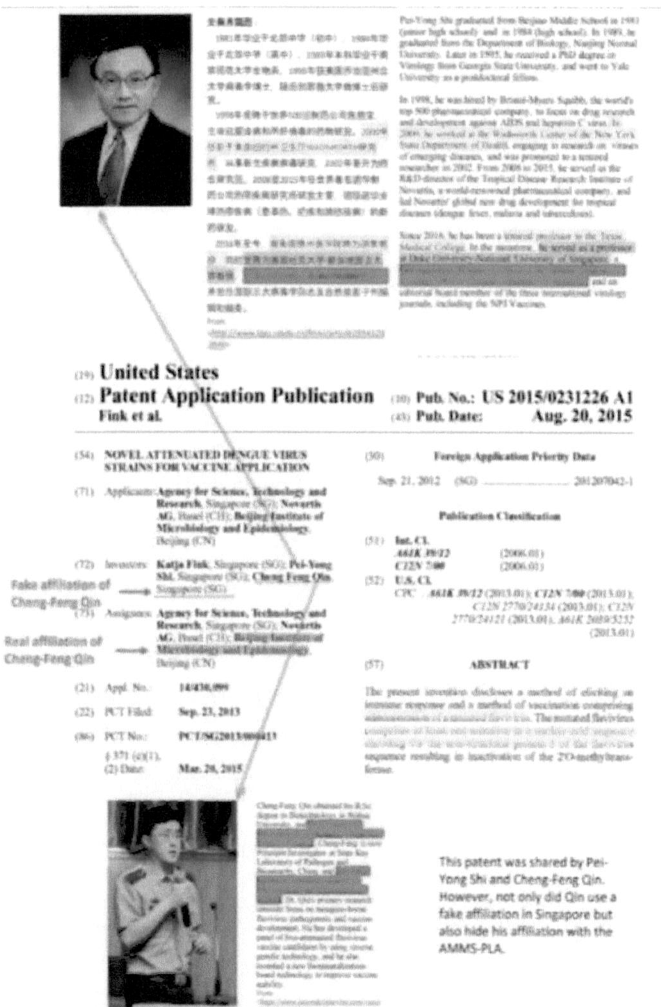

Source: An Annymous E-mail Account

IX

From the Magician's Hat

"So, What is the book's progress" Darpan called me on 3rd December 2021 and asked without delay.

"I think I found the 'Smoking Gun,' I told him.

"Which Smoking Gun?" He asked

"Do you remember we talked about the book written by PLA scientists, which U.S state departments later obtained?" I asked him

"Yes, So What?" He asked

"I found the scanned copy of that book in PDF format," I said.

"Oh My God! Are you sure?" He asked to confirm.

"Yes, I have confirmed!" I replied.

"That is Great, Brilliant! But how did you get that? Did you purchase it from somewhere?" He asked.

"No, I do not know if this book is available to purchase right now, but it was previously available on amazon.com, but now it is unavailable. And I got it from Magician's Hat," I laughed and told him.

“That is a great breakthrough! I am now being more eager to read this book.” He said.

“You will not regret it!” I replied and ended the call.

In August 2015, a book Titled ‘The Unnatural Origin of SARS and New Species of Man-Made Viruses as Genetic Bioweapons Book’ was published in the Chinese language. The authors of that book were Li Feng And Xu Dezhong. Xu Dehzong had also published an article on https://pubmed.ncbi.nlm.nih.gov/24985597/#affiliation-1 in 2014. Unfortunately, the link to the full article is diverting to some other website. Ping Shi Medical Publishing House published the book. Du Jing.

The book was also available on Amazon!

Image Source Amazon

The Team of Editors,

The division of labor, units, and titles of the members of the editorial board

Editor-in-chief:

Xu Dezhong, Professor, Department of Military Epidemiology, Fourth Military Medical University; INCLEN, CEU, Director

Li Feng, Deputy Director, Bureau of Epidemic Prevention, Ministry of Health, General Logistics Department; Vice President, China Association for Health Promotion and Education

Deputy Editor:

Wang Anhui, Deputy Director, Associate Professor, Department of Military Epidemiology, Fourth Military Medical University

Li Guanglin, Associate Professor, School of Life Sciences, Shaanxi Normal University

Zhang Lei, Associate Professor, Department of Military Epidemiology, Fourth Military Medical University

Wu Xiuhua, Director, Chief Physician, Department of Gastroenterology, Shaanxi Armed Police Hospital

Editors:

Duan Guangcai, Professor, Department of Epidemiology, School of Public Health, Zhengzhou University

Yang Ruifu, Director, Researcher, Laboratory of Analytical Microbiology, Institute of Microbiology and Epidemiology, Academy of Military Medical Sciences

Song Yajun, Deputy Director, Researcher, Laboratory of Analytical Microbiology, Institute of Microbiology and Epidemiology, Academy of Military Medical Sciences

Zhang Jingxia, Department of Military Epidemiology, Fourth Military Medical University, Senior Experimentalist

Tang Xiaofeng, Deputy Chief Technician, Outpatient Department, Fourth Military Medical University

Wang Bo, Associate Professor, Department of Military Epidemiology, Fourth Military Medical University

Sun Huimin, Post-doctorate, Department of Military Epidemiology, Fourth Military Medical University

Li Duan, Lecturer, Department of Military Epidemiology, Fourth Military Medical University

editor:

Su Haixia, Associate Professor, Department of Military Epidemiology, Fourth Military Medical University

Wang Yingfang, Lecturer, Department of Public Health, School of Medicine, Henan University of Science and Technology

Xu Rui, Department of Military Epidemiology, Fourth Military Medical University, PhD

Zhao Ningning, Department of Military Epidemiology, Fourth Military Medical University, Assistant Experimentalist

According to that book, it was clear that the Chinese military scientists discussed weaponizing SARS Coronaviruses. The book outlines China's progress in the research field of biowarfare.

From Here Onwards, most of this chapter is the English translation of the Book published in Chinese under the title, '*The unnatural origin of sars and new species of man-made viruses as genetic bioweapons*' by Li Feng & Xu Dezhong. The translation has been mentioned not to change the real meaning published in Chinese! So no grammatical corrections were applied from here onwards to maintain the originality of the translation.

Page No. 6, second paragraph of the book, explains that,

'Paying attention to the painstaking efforts: In the day and night of thinking, finally one night, my mind suddenly flashed a description that has been studied many times, S A R S-

A picture in the CoV molecular evolution paper, and associate it with "reverse evolution". Suddenly, it became clear: I found it dreaming for many years, the "golden key" that opened the door to the origin of SARS-"reverse evolution" theory and technology! After a long period of time in nature. However, new viruses that have evolved to adapt to animals that are closely related to humans, and then gradually adapt to humans and spread in their populations. It must be based on "forward evolution"; on the contrary, if it is a new virus that has an unnatural origin and has not undergone this natural adaptive evolution process. In the epidemic, it is bound to be unable to adapt to the new host of humans and "reverse evolution" appears, and finally leaves the crowd!

Second, the relationship between the origin of the SARS virus and the genetic weapons of the new human-to-human virus. This book is based on epidemiology, infectious disease.

A large number of evidences such as science, molecular virology, molecular biology and molecular evolution have determined that SARS-CoV is of unnatural origin, that is, man-made origin. At present, various terrorist activities are rampant, and there is no exception to bioterrorism with the rapid development of biotechnology.

With development, the development methods and types of biological weapons and genetic weapons are rapidly updated. Therefore, the human origin of the SARS virus cannot be excluded.

Except it is a new (contemporary) genetic weapon developed by terrorists. Unexpectedly, this assumption was confirmed in the summer of 2013.

Reality. At that time, when I checked the literature, I stumbled upon Michael Annis, Air Force Colonel, Air Force

War Academy, Air Force University of a certain country.'

Translation of fragment of the second paragraph from Page No. 6 of 'The Unnatural Origin of SARS and New Species of Man-Made Viruses as Genetic Bioweapons'

Extraction of page no. 7 of the book states, *"Before applying the theory of "reverse evolution" to crack the unnatural origin of SARS-CoV, we will first apply the epidemic of infectious diseases in my country. The epidemiological theory, combined with the philosophical thinking of seeing the essence through the phenomenon, reveals the unnaturalness of SARS in epidemiology and clinical medicine. The truth of the origin, and pointed out the connection between it and the "human-based new virus genetic weapon".*

Page No. 8, *"It is true that the editorial board has invited well-known experts in the fields of digital molecular biology, molecular evolution, and genetic weapons to participate."- March 11, 2015*

In mid-May 2003, the then Director Qi Xiaoqiu (middle) of the Department of Disease Control and Prevention of the Ministry of Health and Professor Xu Dezhong conducted the second "Focus on SARS Prevention and Control" topic. Preparation scene of Interview

Translation of Book Content,

content

Excerpts and why she is so familiar with the origin and disappearance of SARS but not fully elucidated... (222)

Appendix 5 "Chinese Rhinolophus is a storage place for SARS-like virus (SL-CoV) but not SARS-CoV (SARS-CoV)

Main --- Criticism of Ge article in "Nature" magazine, R. sinicus should be reservoir of SL-CoV

but not SARS-CoV:critical comments on Ge et al. 's paper in Nature" reprinted ... (228)

Reproduced in Appendix 6 "The Abnormal Distribution of Human Infection with H7N9 Avian Influenza and the Possible Origin of Abnormality"...... (235)

We will discuss this book Page by Page with important and insightful content!

Section 1 Biological Agents and Biological Weapons

Chapter – 2 Introduction to Biological Weapon.

Biological weapon (biological weapon) refers to the general term for biological warfare agents and their loading and release devices.

It is composed of a biological warfare agent, projectile body, release device, propulsion device, timing device, and blasting device.

1. The characteristics of biological warfare agents

Currently, there are multiple classification methods for biological warfare agents. According to biological characteristics, warfare agents were once classified into bacterial, viral, and fungal. There are four types of sex and toxin warfare agents. Bacterial warfare agents include Bacillus anthracis, Yersinia pestis, Francis tularensis, Brandt Bacillus coli, Coxiella baekeri, etc. Viral warfare agents include variola virus, Oriental equine encephalitis virus, Venezuelan equine encephalitis virus, Yellow fever virus, etc., fungal warfare agents include coccidioidomycetes, histoplasma capsulatus, etc., toxin warfare agents include botulinum toxin, ricin, Vegetarian, staphylococcal enterotoxin B, etc.

Although many kinds of pathogenic microorganisms and toxins can cause human diseases or poisoning, only a few can be used as biological organisms. Warfare agents are used in biological weapons. Pathogenic microorganisms or toxins that can be used as biological warfare agents on a large scale need to be basically or completely meet the following conditions: Can be easily mass-produced; Can cause death or disability within the range of castable amount; Highly contagious (Except toxins); produce aerosols of suitable size particles; easy to spread; after production, storage, weaponization and release, etc. It is stable under conditions and maintains virulence; the enemy is sensitive and oneself has effective means of protection. Of course, with the development of biological weapons, the conditions for being a biological warfare agent will change accordingly.

1. Easy to produce. Many bacteria and viruses can be mass-produced by modern fermentation and culture techniques respectively; some toxins, such as Ricin has a wide range of sources and mature extraction technology, which can also meet the needs of war agent production. Although there are still many pathogenic microorganisms. The growth is slow, the titer is not high, and certain toxins are limited in nature, but they can be solved by modern biotechnology in the future.

2. The use of lethality, disability, and the length of the incubation period are divided into two types: usually 10% as the boundary, the lower one is a disabling warfare agent, and the high one is a lethal warfare agent. The former will deprive the enemy troops of war. Fighting power increases the burden of the enemy's medical support; the latter

directly kills a large number of enemy troops. Venezuelan equine encephalitis (VEE) disease.

Page No. 51

How to release biological warfare agents

'Biological warfare agents can be cast in a wet or dry state. As mentioned above, dry cast is ideal, but the drying of warfare agents requires complex technology. The release of warfare agents can be through spray devices, explosive devices, artificial contamination of food and water, and the release of infectious agents, vectors and other methods. The most common method is aerosol release. As mentioned above, it is a better way to install the spray device on the vehicle for aerosol delivery. Its casting path will be a linear pollution zone is formed, which is called line source release. This method is generally driven against the wind to cast, it feels infect the crowd within a certain range of the wind direction. The extent of pollution depends on wind speed, wind direction, meteorological conditions, topography and vegetation, and the natural physical characteristics and other factors. Stable warfare agents (such as anthrax) can contaminate a range of 20 km downwind. A WHO study table It is clear that under ideal weather conditions, if 50 kg of aerosolized anthrax spores are sprayed on the upper wind side of a residential area by plane.'

Page No. 156-157 of Book: The Unnatural Origin of SARS and New Species of Man-Made Viruses as Genetic Bioweapons

In Guangdong, my country, the first case occurred from November 16, 2002 to January 31, 2003. There were 194 cases of onset in two and a half months, scattered in 6 urban areas, with slow transmission and mild symptoms. The case-fatality rate was low; and there were 1037 cases in the next two months, mainly in Guangzhou (Figure 6-2), and it was a hospital-family outbreak model with critical illness and more super-spreaders. During the SARS epidemic, Guangdong Province

Cases have been reported in 15 of 21 prefecture-level cities. The cases were mainly concentrated in the five cities of Foshan, Guangzhou, Shenzhen, Zhongshan, and Jiangmen in the Pearl River Delta, accounting for 96.03% of the total number of cases. Among them, Guangzhou had the largest number of reported cases, accounting for 86.04% of the total number of cases.

What is more intriguing is the regional distribution characteristics of SARS in the early and mid-term epidemics in Guangdong Province. Science magazine gave a special description of this: From the first case to the index case in Dongguan on March 10, 2003 (that is, my country) In the 3 inland epidemic periods, the early to late period). During this period, the cases all occurred in several cities to the west of Guangzhou and Shenzhen to the south.

There was no case (Figure 6-2); for the onset of Heyuan, the paper specifically explained: its index case was infected in Shenzhen, and there were no patients in Heyuan at that time. The distribution of such abnormal regions is thought-provoking: because the cities to the east and north of Guangzhou are not significantly different from the west and south in terms of social and natural environment.

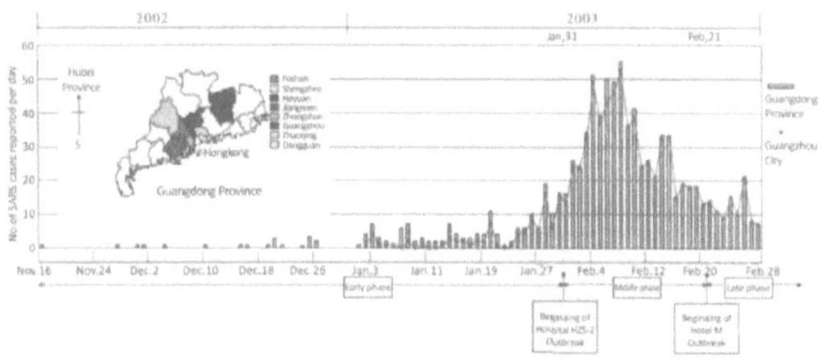

图 6-2　广东省 SARS 发病时间和地区分布(2002 年 11 月—2003 年 2 月)

(引自:Chinese SARS Molecular Epidemiology Consortium. Science, 2004, 303: 1666–1669.)

Image Source: The Unnatural Origin of SARS and New Species of Man-Made Viruses as Genetic Bioweapons Book by Xu Dezhong and Li Feng

Further, the authors write,

In this regard, the author explained in the paper as follows: Due to the economic development of the Pearl River Basin in the early 1970s, the residents' eating habits changed and they began to introduce foreign animal food. This seems to have some truth, but it cannot explain the following questions: 1.The habit of eating civet cats has been around for decades, but why did it become popular in 2002; 2. This habit is all around Guangzhou, why is it popular early? The middle stage is limited to the west and south of Guangzhou; 3. The civets that can separate **SA R S-Co V are only found in Guangzhou and Shenzhen wild animal markets**, but the surrounding rural farms and other provinces have none. Therefore, we believe that the deep-seated reason (internal essence) reflected by this strange distribution (external manifestation) is likely to be the infection of the **civet cat and/or the input of the infected**

civet cat, that is, the source of infection is the animal kingdom And (or) the crowd is unnatural, in other words artificial. This view put forward by epidemiological analysis is consistent with the "reverse evolution" process in the development of SA RS-CoV system involved in many places in this book.

Page No. 157 Paragraph No. 2

In epidemiology, when analyzing the causes or mechanisms of epidemics or outbreaks, corroboration is emphasized, and its role is like an important or even key link in the evidence chain of judicial cases. **From March 25 to the end of April, 2004, a laboratory infection occurred in my country. A total of 9 SARS patients were diagnosed in Beijing and Anhui, and 1 case died.** There are continuations and super spreaders in the outbreak.

This laboratory infection, SARS-CoV has strong toxicity and transmission power, so its epidemic intensity is high, and its epidemiological characteristics it is obviously different from the outbreak in Guangzhou in January of the same year, but it is very consistent with the epidemic in 2002-2003. From the perspective of the SARS epidemic, **this laboratory infection was neither a continuation of the SARS epidemic from 2002 to 2003, nor was it a continuation of the SARS epidemic in Guangzhou that occurred in the previous 3 months.** The outbreak has nothing to do with it, it should be independent.

However, the peculiar manifestation of the epidemic distribution of this laboratory infection is not accidental; it is exposing the tortuous evolutionary process that SARS-CoV has experienced during the epidemic: due to its unnatural origin, it is fierce, but because it is not adapted to the population, it can only In response to "reverse evolution", the virulence and transmission power declined rapidly. Therefore, there were no recurring cases of the outbreak in Guangzhou from December 2003 to January 2004.

example. However, this kind of "reverse evolution" only occurs in the SARS-CoV strains that are prevalent in the population; and the strains stored in the laboratory in the first half of 2003 still maintain the original virulence and transmission, so laboratory infections have occurred. After the outbreak in Guangzhou from March to April 2004, it still showed the violent state of the epidemic from 2002 to 2003. As a result, it strongly corroborates or disproves the "reverse evolution" and unnatural origin of SARS-CoV.

Page No. 159

Undoubtedly, the reasons are as described above. The SARS-CoV that caused this outbreak has been recurrent, with low virulence and no ability to retransmit. The virus in the patient's body was quickly cleared. From this, readers can see that, just like the epidemiological characteristics of epidemic distribution and the "phenomenon and essence" relationship between the three links, the clinical type and severity of the disease are related to pathogen variation and virulence. The relationship between "phenomenon and essence" is also present: the "phenomenon" of the Guangzhou outbreak has changed clinical types, mild symptoms, and good prognosis, reflecting the "essence" of SARS-CoV that has been "reversely evolved" at this time! Of course, as pointed out in the previous chapter, further analysis can reveal that the "reverse evolution" of SARS-CoV during the population epidemic is a reflection of its "unnatural origin"! Thus, in the field of clinical science, it provides strong evidence for the conclusion that SARS-CoV has a "reverse evolution" and an unnatural origin.

无疑，其原因已如上所述，造成此次爆发的 SARS-CoV 已经返祖性变异，毒力低，无续发传播能力，患者体内的病毒迅速被清除。由此读者可见，和流行病学上流行分布特征与三个环节之间的"现象与本质"关系一样，临床类型、病情轻重的特征与病原体变异、毒力高低也同样呈现"现象与本质"关系；广州爆发临床类型改变、症状轻、预后好的"现象"，反映了此时的 SARS-CoV 已经"逆向进化"之"本质"！当然，如前章指出的，进一步分析可揭示 SARS-CoV 在人群流行期间出现的"逆向进化"又是其"非自然起源"之反映！从而在临床学领域方面，为 SARS-CoV 存在"逆向进化"且非自然起源之论断提供了有力证据。

表 6-2　2003 年底至 2004 年初广州爆发病例一般特征

编号	姓名	性别	年龄（岁）	职业	住址	流行病学史			
						同类患者接触史	野生动物接触史	外出史	医院就诊史
1	罗某	男	32	独立制片人	广州番禺区	无	无	无	无
2	张某	女	20	A 餐厅服务员	广州越秀区	无	有	无	无
3#	杨某	男	35	发廊个体户	广州东山区	无	*	无	无
4	刘某	男	40	医院院长	广州海珠区	*	无	无	*

注：#代表病例 3,4 分别在发病前 7,8 天曾到 A 餐厅（经营果子狸）及其隔壁 B 餐厅就餐；
*代表情况不明

Page No. 159 of The Unnatural Origin of SARS and New Species of Man-Made Viruses as Genetic Bioweapons Book

Page No. 125 – 126: The Unnatural Origin of SARS and New Species of Man-Made Viruses as Genetic Bioweapons Book

(2) Strategies to crack contemporary genetic weapons

In response to "super-war" weapons, we should not associate them with "war", that is, in peacetime, especially when our country suffers

When containment, national sovereignty is being or has been threatened by terrorists, diseases caused by unknown causes or new species of pathogens occur

When the disease is spreading on a large scale, one should consider whether there is the possibility of being attacked by contemporary genetic weapons.

Therefore, the strategy of cracking contemporary genetic weapons should be considered from two aspects: academic and strategic situation.

1. The academic strategy for deciphering contemporary genetic weapons has been mentioned in this chapter, 'During the Korean War, it was determined that the enemy launched a germ warfare

The principle is: the epidemic caused by it does not conform to its natural history! It should be emphasized here that this principle is still applicable to this day and is to identify any type

The "gold standard" of genetic weapons! '

The author thinks that although there is a "gold standard", the philosophical thinking of "specific analysis of specific situations" must be applied in practice. which is with the development of the times and technology, its meaning should be expanded. However, in order to cope with the current situation, it is necessary to give a definition and Give a more detailed description of its strategy.

Definition: The gold standard to recognize contemporary gene weapons are the emergence and/or epidemic process of new diseases that suddenly appear and the evolutionary course of their pathogens or disease-causing genes none of them correspond to their respective natural history!

Strategy: The connotation should include at least the following points.

(1) The pathogen or disease-causing gene that caused the occurrence or epidemic of the disease is a new artificial variety, so it does not conform to the natural evolutionary history of the same species. **Taking SARS-CoV as an example, the current academic circles believe that the Bt-**

SLCoV Hp3 (DQ071615) strain is the common first, and the average time between evolution with SARS-CoV is 4.08 years. From this, it can be inferred that such a short evolutionary time. Time does not conform to the law of natural evolution, that is, it is of unnatural origin.

(2) There are no direct ancestors of pathogens or disease-causing genes that caused the occurrence or epidemic of the disease in nature or in the population. Because it is a new artificial species produced in the laboratory, its direct ancestors cannot be stored exclusively in nature or in natural populations. The ground is in the laboratory of terrorists.

SARS non-natural origin (S>____

Human-to-human new virus genetic weapon

It can be seen that the reason why the direct ancestor of SARS-CoV has not been discovered so far. If you can't find it in nature see the direct ancestor of h-H7N9 AIV, which is the main reason for identifying it as an "ecological genetic weapon of a new species of human-to-human virus".

(3) The way of transmission of pathogens or disease-causing genes that caused the occurrence or epidemic of the disease is abnormal. **This book has discussed that SARS was first launched in the Pearl River Delta, which is close to the sea; civet cats as a source of infection only exist in Guangzhou and Shenzhen markets, and wild animals in the market can circulate privately with foreign countries.**

(4) The disease or epidemic caused by it does not conform to the epidemiology and clinical medicine rules of the disease caused by the naturally evolved pathogen or pathogenic gene. It has been mentioned in many places in this(***The Unnatural Origin of SARS and New Species of Man-Made Viruses as Genetic Bioweapons***) book that the epidemic process and characteristics of SARS and the

changes in clinical characteristics during the epidemic process and similar infectious diseases, and even acute human infectious diseases so far are obviously abnormal. It has also been discussed that many epidemiology of h-H7N9 AI. **Therefore, it cannot be excluded that it is of unnatural origin because of the abnormality of various A-type h-Al that has appeared.**

In the years before the SARS epidemic in 2002, the people of our country should remember that there have been several terrorist incidents that blatantly threatened our sovereignty. It is difficult to rule out the hidden connection between these terrorist incidents and the SARS epidemic.

Page No. 128

Subsequently, 3 months after the outbreak in Guangzhou, **there was a Beijing-Anhui SARS laboratory infection, with a total of 9 cases. The epidemiological and clinical characteristics are exactly the same as the epidemic from November 2002 to July 2003: there is a super spreader, and the condition is serious: 1 person died; however, it is significantly different from the Guangzhou outbreak. This is another remarkable anomaly in epidemiology and clinical medicine (actually including virology and molecular evolution)!**

非典非自然起源和
人制人新种病毒基因武器

2. 同时应用"演绎法"和"归纳法"的推理思维　我们可将上述各种各类零碎、单篇，甚至片面和局限的信息，结合生物技术之发展趋向和风云变幻之恐怖态势，在相关学术理论指导下，进行整理并对其采取"演绎法"和"归纳法"的推理分析：去粗取精、去伪存真、由此及彼、由表及里，获取其真实之意图，揭示其当代基因武器之本质。

揭示 SARS-CoV 非自然起源，甚至为当代基因武器本质的思维过程，是此种方法之最好实例。

(1)对 SARS 流行，首先映入眼帘的是，猛烈席卷全球一年多后，为何无影无踪？对此，可能普通大众也将如此发问。实际，此问是 SARS 在流行病学上最关键之反常！人类传染病史上，尚未有任何其他传染病如此。所以，可推理：SARS 流行不符合人类传染病自然史，SARS 流行有异常！进一步质疑：引起其流行之病原体 SARS-CoV 产生或起源可能有异常。

(2)SARS 在 2002 年 11 月至 2003 年 7 月连续流行后，为何相隔数月后于 2003 年12 月至 2004 年 1 月发生仅 4 个病例之广州爆发，而后者在流行病学和临床医学特征和前者明显不同。此问又进一步支持上述质疑！

Page No. 128 From Book, The Unnatural Origin ...

SARS-CoV is not adapted to humans, and has produced "reverse evolution", and finally left humans, proving that it is an unnatural evolution, and it is derived from bat SL-CoV or other similar SL-CoV through laboratory genes Developed by methods such as transformation and preliminary animal adaptation-passage test. The above-mentioned values, evidence, and events used for reasoning are all extracted one by one from the bulletins and documents published in domestic and foreign magazines and international organizations after the occurrence of SARS; then the relevant sorting is carried out purposefully, and various reasoning and comprehensive analysis are considered at the same time. , And finally get the above points.

Even though SARS in the Middle East appeared the latest, its pathogens MERS-C o V and SARS-C o V belong to the P coronavirus (p-CoV), but epidemiological, virological and molecular evolution studies have initially confirmed that a certain local Some species of bats and camels are the source of infection, and may even be reservoirs; moreover, from the current global epidemiological distribution, it can be preliminarily believed that the Middle East is the natural focus of SARS in the Middle East.

From this, it can be determined that, throughout the ages, only SARS-CoV has not found direct ancestors (h-H7N9 AIV has yet to be discovered) and reservoir hosts in nature.

Page No.- 136

Initially, Hong Kong scientists detected human SA R S-CoV homology in civet cats

As for the extremely high coronavirus (Citrus civet SARS-CoV), researchers began to explore the possibility of civet cats as the source of infection of SARS and even the reservoir of SARS-CoV. However, many studies have only confirmed that during the SARS epidemic, civet cats can be used as a source of SARS infection; after the end of the epidemic, SARS-CoV has not been isolated in civet cats. Therefore, as mentioned above, we refer to the civet cat as the "specific source of infection" or "temporary source of infection" of SARS, and it is impossible to store it as host.

Since then, scholars searching for the source and reservoir of SARS-CoV have focused their attention on bats. However, the researchers collected 905 bats from Guangzhou and surrounding areas at Guangzhou and Guangzhou Wildlife Markets from 2004 to 2005. They also collected 8 13 bat swabs, 5 24 serums, and lungs. Organize 853 specimens and 853 rectal stool specimens. The test results showed that none of the 3 0 4 3 specimens of 9 kinds of 905 bats were separated from SARS-C o V or S L-C o V. However, in 2005, some scholars found that the nucleotide sequence of the coronavirus genome isolated from bats had a high homology of 92% with the human SARS-C o V genome, the homology of the S gene was 64%, and the homology of P and E The homology of, M and N proteins is between 96% and 100%; molecular evolution analysis shows that SARS-like-CoV (Bt-SL CoV) in bats is the same as human and civet SARS-CoV Evolutionary branch. Subsequently, researchers isolated Bt-SLCoV from common long-winged bats, big-footed myotis and Chinese chrysanthemum-headed bats, which have high homology with SARS-CoV genes. **In particular, evolutionary analysis showed that the Bt-SL CoV Rp3 (DQ071615) strain of Chinese chrysanthemum bat is the common ancestor of SARS-CoV, but it is not a direct ancestor. There is a 4.08-year-old between the two Evolution time interval.**

Although the current molecular detection technology is advanced and the genetic evolution analysis method is powerful, scientists around the world have yet to reveal the direct origin and storage host of SARS-CoV. There are many reasons for pondering. One of them is a more detailed introduction for the first chapter of this book: the different understandings of theories of natural focal diseases and zoonotic diseases in the epidemiological academic circles of the East and the West. The concepts and functions of natural foci diseases have been introduced in more detail in this chapter. **Since SARS has not found its direct ancestor and reservoir in nature, it cannot be a natural foci disease.**

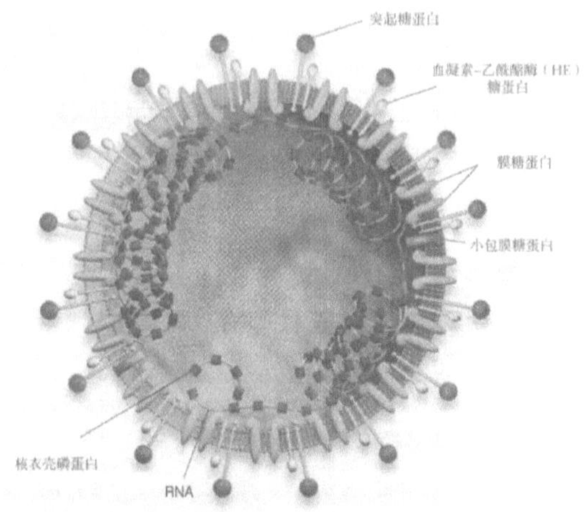

图 7-1　冠状病毒形态和基因结构的示意图

Spike(S)glycoprotein：突起糖蛋白；hemagglutinin-acetylesterase(HE)-glycoprotein：血凝素-乙酰酯酶(HE)糖蛋白；Membrane(M)glycoprotein：膜糖蛋白；Small envelope(E)glycoprotein：小包膜糖蛋白；Nucleocapsid(N)phosphoprotein：核衣壳磷蛋白
[引自：Holmes KV. SARS-associated Corona Virus. N Engl J Med，2003，348(20)：1948-1951]

Chapter VI Evidence of the Epidemic and Clinical Features of the Non-natural Origin of the SARS Virus............. (151) of the book states,

第六章　非典病毒非自然起源的流行特性和临床特性之证据

从 2002 年底 SARS 流行开始至今，尽管已经过去 10 多年，但整个医学界对 SARS-CoV 的起源仍未知晓。这在高科技如此发展的今天难以想象，在近代人类传染病流行史上实属罕见。诚然，国内外无数学者对其基因结构及其变异与跨宿主进化进行了全面深入的研究，但是仍然没有发现 SARS-CoV 的真正起源。究其原因，主要在于这些研究者多数来自实验医学领域，没有从宏观和微观结合上，没有从同一疾病发展前后历史比较上和多病种相互比较上，以及未在创新思路上探讨 SARS-CoV 起源的实质。

最近，国内学者徐德忠教授及其团队通过多年苦心研究后认为，自然界根本不存在 SARS-CoV 的直接祖先和贮存宿主，其是由 Bt-SLCoV 经“非自然（如基因改造结合动物传代适应试验）进化（unnatural evolution，UE）”方式产生。该病毒在短暂流行后不久即在自然界和人群中消失，故他们将 SARS-CoV 称为“过客病毒（passenger virus）”。这实际上反映出人类已进入经基因改造技术和群体动物传代试验相结合等方式、将某些低级野生动物病毒人工制成“新品种人类病毒（即非原有病原体之基因改造）”，并引起一国或多国乃至全球性流行的新时代，故应引起各国科学家和公众的足够重视。因此，本章从疾病的分布和临床特征是否反常的角度出发，为进一步详细阐明 SARS-CoV 的非自然起源提供证据。

Chapter VI Evidence of the Epidemic and Clinical Features of the Non-natural Origin of the SARS Virus............. (151) Source: The Unnatural Origin of SARS and New Species of Man-Made Viruses as Genetic Bioweapons Book

Recently, after years of painstaking research, a domestic scholar Professor Xu Dezhong and his team believe that there is no direct ancestor and reservoir of SARS-CoV in nature. It is produced by Bt-SLCoV through “unnatural (such as genetic modification combined with animal passage adaptation test) evolution (unnatural evolution, UE) n method. The virus disappeared in nature and people shortly after a short epidemic, so they called SARS-CoV "passenger virus (passenger virus) M." This actually reflects that human beings have made some low-level wild animal viruses artificially into "new species of human viruses (that is, genetic modification of non-original pathogens)" through the combination of genetic modification technology and group animal passaging experiments. And cause one or more countries and even global. The new era of popularity should arouse sufficient attention from scientists and the public from all over the world. Therefore, this chapter provides evidence for further elucidating the unnatural origin of SARS-CoV from the perspective of the distribution of the disease and whether the clinical features are abnormal.

If the epidemiological characteristics of the new infectious disease are significantly different from those of the same infectious disease or even abnormal, it cannot be ruled out that it is an unnatural epidemic, and its pathogen is of unnatural origin; if it is consistent with history Human infectious diseases are abnormal, and combined with evidence from other disciplines or a chain of evidence has been formed, **it is sufficient to judge that the pathogen is of unnatural origin!**

Page No. 181

Here, before making the hypothesis on the origin and reverse evolution of SARS-CoV, it is necessary for the author to briefly review the emergence, evolution and prevalence of SARS-CoV and the main characteristics of the research process.

1. The earliest ancestor of SARS-CoV is most likely Bt-SL CoV (Rp3 strain), and the common ancestor of the two is about 4.5 years before the SARS outbreak. If only such a short period of natural evolution, it does not conform to the law of virus evolution; in other words, such a short period of evolution must be unnatural evolution.

2. The direct ancestor and reservoir of SARS-CoV have not been found in animals so far; there is no precedent in the history of human infectious diseases and research history.

3. Except for the Guangzhou outbreak and laboratory infections from December 2003 to January 2004, there was no vaccine prevention. The SARS epidemic lasted only 8-9 months; the time points of the outbreak and disappearance were sudden; it is also unprecedented.

4. The epidemiological and clinical characteristics of SARS and other human coronavirus diseases are very different, that is, they do not conform to the laws of other human coronavirus diseases.

5. The general laws of the evolution and epidemic of SARS and SARS-CoV and emerging infectious diseases in recent years are abnormal.

6. There are many characteristics of reverse evolution in the evolution of SARS-CoV, which is unique in the evolutionary history of pathogens in the epidemic of human viral diseases so far.

7. When SARS became an epidemic, it took more than 8 months (November 2002 to July 2003) in 29 countries and places around the world a total of 8,098 cases occurred in the district, but since then, apart from the Guangzhou outbreak and laboratory infections from December 2003 to January 2004, humans and animals have not been infected for more than 10 years; and the natural history of the epidemic of all human viral diseases is abnormal.

Page No. 182-183

All these have proved that the origin of SARS-CoV is extraordinary and does not have the characteristics of natural evolution. It is an unnatural origin, that is, SARS-CoV is genetically modified from Bt SL-CoV (Rp3 strain)

or similar virus strains combined with a population adaptation test And other methods; otherwise, the above points cannot be explained.

Regarding this picture, in addition to the annotations, there is also a point of concern that people need to discuss, that is, methods such as genetic modification combined with population adaptation experiments. The same problem has been mentioned in many places in this book. A detailed introduction is given in Chapter 4 of this book. It is worth mentioning that in the past one or two decades, various media and channels have discussed that humans have long mastered this or similar methods. Therefore, there is no doubt that some terrorists have initially or even mastered the technology of unnaturally transforming bat Bt SL-CoV Rp3 or similar viruses into SARS-CoV around 2000.

Although the individual links and details of Figure 7-6 need to be supplemented and modified, the following questions can be reasonably explained in general

And facts:

1. Why after February 2004, SARS has no more cases and epidemics in the population except for laboratory infections.

2. Why hasn't the direct ancestor and reservoir of SARS-CoV been found in nature so far?

3. Why the SARS clinical and epidemiological characteristics are quite different between the Guangzhou outbreak from December 2003 to January 2004 and the population epidemic from November 2002 to July 2003.

4. Why is the genetic relationship between humans and civets SARS-CoV in the Guangzhou outbreak from December 2003 to January 2004 closer to the early strains of the population epidemic from November 2002 to July

2003, rather than late?

5. Why is the variation of the important AA site in the S1 region where SARS-CoV binds to the host receptor, the variation or deletion of ORF8 29-nt, and the stop codon in SUD, etc.

(Sun Huimin, Zhang Lei, Xu Dezhong, Zhao Ningning and Xu Dezhong)

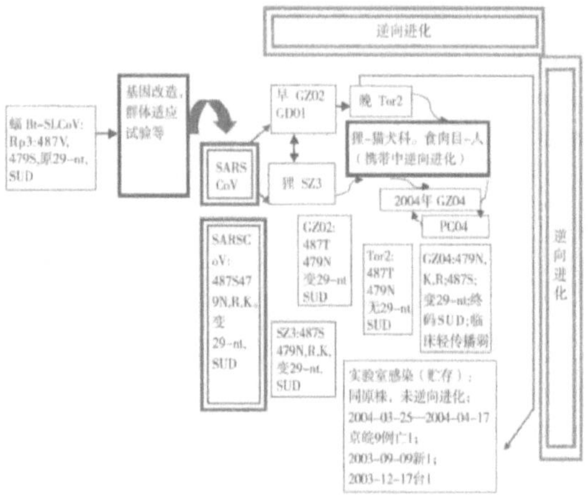

图 7–6 "SARS–CoV 的起源和进化假设"的框架图

1. 图中框,总体分二层,上层为 SARS–CoV 的起源和进化流程,下层除"实验室感染"外,为上层框的内容或毒株特性之描述。2. 487、479:SARS–CoV 或 Bt SL–CoV S1 区 487,479 位 AA,数字前或紧随其后的大写英文字母为 AA 之种类。3. 原29–nt,变 29–nt,无 29–nt:分别代表 Bt SL–CoV 或 SARS–CoV ORF8 特征性 29–nt 序列:CCAATACATTACTATTCG-GACTGGTTTAT(nt. 27866–27894),CCTACTGGTTACCAACC TGAATGGAATAT(nt. 27869–27897)和特征性 29–nt 序列缺

Page No. 182

<u>Page No. 232-233</u>

Research on the origin of SARS coronavirus (SARSCoV) and SARS-like coronavirus (SL-CoV)

New infectious diseases caused by viruses are very important. Recently, Ge X Y et al. proposed that "Chinese chrysanthemum bats (C h in e s e h o r s e s h o ebats)

are the natural reservoir (reservoir) of SARS CoV" (1). **But this conclusion is not correct, it should be "Chinese chrysanthemum bat is the storage host of SL-CoV" [2"7]. There are several methodological problems in this paper, which lead to some errors and wrong conclusion.**

Page No. 233-234

The key amino acid positions 479 and 487 of SARS-CoV or SL-CoV carried by bats, humans and civet cats are quite different (Table 1). More importantly, Ge XY and others did not compare the human/civet cat isolates that broke out in Guangzhou. The AA 4 7 9 and 4 8 7 positions of these strains are not only different from Rs3367 and RsSHC014, but also different. Human SARS-CoVs from 2002 to 2003 [2*4*6>8*10'11]. This reflects that there is no essential difference between the two SL-CoV (Rs3367 and RsSHC014) and the previous ones.

In their paper, 9 years after the SARS epidemic, they discovered that the Chinese chrysanthemum bats carried the SL-C that "has a typical coronavirus morphology and has 99.9% homology with the R s 3 3 6 7 sequence". o V. Based on this, they came to the conclusion that "Chinese chrysanthemum bat is the natural reservoir of SARS-CoV" [1]. However, during the SARS epidemic and in the following years, no R s 3 3 6 7 or R s S H C 0 14 strains were found [2_71. Therefore, Ge X Y and others cannot rule out the possibility that their discovery is only the natural evolution of SL-CoV in Chinese chrysanthemum bats in recent years.

Seventh, there is a lack of alignment of specific 29-nucleotide (specific 29-nt). Characteristic 2 9-n t (B t-S L-C o V is n t. 2 7 8 6 6-2 7 8 9 4 C C A A T A C A T T A C T A T T C G G A C T G G T T T A T, while human/ fruit/the civet SARS-CoV is nt. 27869 -27897 CCTACTGGTTACCAACCTGAATGGAATAT) is closely related to the pathogenicity, transmission power and cross-species transmission of SARSCoV. Since the characteristic 29-nt of R s 3 3 6 7 and R s SHC 0 1 4 is consistent with SL-CoV, but different from SARS-CoV (Table 1), these two types of strains should be included in SL-CoV.

We are not without worries, the research conclusions of Ge XY and others will not only have a negative impact on the scientific community, but also hinder the scientific research on the true origin of SARS-CoV.

Page No.-180-181

SARS-CoV lacks new adaptive mutations: In the new host, SARS-CoV has poor new selectivity conditions, not seen new adaptive mutation; on the contrary, after entering the human population, it quickly lost the "characteristic 29 nt sequence" obtained from its ancestor strain, a series of reverse evolutions followed. In fact, it can be deduced inversely. It is very likely that this "characteristic 29 nt sequence (variation) 29 nt)", not due to natural evolution!

As a result, one year after the epidemic, the virulence and transmission of SARS-CoV dropped drastically, and soon disappeared from humans and nature. From the perspective of natural evolutionary history, leaving the crowd and/or returning to its ancestral state is the only and best ending.

If it is highly generalized, one sentence can be used to point out the reason for the reverse evolution of SARS-CoV: this virus is not suitable for humans at all!

And why is "SARS-CoV totally unsuitable for humans"? The only answer in epidemiology: Because SARS-CoV is not itself natural origin, that is, it was not produced by natural evolution in nature!

关系很近的动物群体内做流行试验或完成此类试验，故人类必然是其非常不适应的新宿主。

（2）生物逆向进化的时间因素：如上指出，“若一个群体迁入一种新环境时间很短，其易保持永久遗传变异”；SARS-CoV 非常类同，其产生后，迅即进入一个全新的陌生环境，故易保持祖先环境产生的永久遗传变异。

（3）SARS-CoV 缺乏新的适应性变异：在新宿主中，SARS-CoV 新的选择性条件差，未见新的适应性变异；相反，其进入人群后，很快丢失了从其祖先株获得的“特征性 29 nt 序列”，继之又出现了一系列的逆向进化。实际上，又可反推，很可能此种“特征性 29 nt 序列（变 29 nt）”，非经自然进化所致！

由此，在流行一年后，SARS-CoV 毒力和传播力剧烈下降，不久即在人类和自然界消失；从自然进化史视之，离开人群和/或回归其祖先状态是其唯一的、最好的结局。

若高度概括，可用一句话点明 SARS-CoV 逆向进化的原因：此病毒完全不适应于人类！而为何“SARS-CoV 完全不适应于人类”？在流行病学上之唯一答案：因 SARS-CoV 为非自然起源，即其不是在自然界由自然进化产生！

二、SARS-CoV 非自然起源及其学术假设

（一）SARS-CoV 非自然起源的主要依据

本章实际上主要围绕 SARS-CoV 的逆向进化和非自然起源展开论述和讨论。在此，在做出“SARS-CoV 的起源和逆向进化假设”前，笔者有必要再简要回顾 SARS-CoV 的出现、进化和流行及其研究过程中的主要特点。

1. SARS-CoV 的最早祖先最可能是 Bt-SL CoV（Rp3 株），两者共同祖先时间约距 SARS 爆发 4.5 年。若仅经如此短时间的自然进化，不符合病毒进化规律；换言之，如此短时间的进化，必为非自然进化。

2. 至今未在动物中找到 SARS-CoV 的直接祖先和贮存宿主；这在人类传染病流行及其研究史中，未见先例。

3. 除 2003 年 12 月至 2004 年 1 月广州爆发和实验室感染外，并无疫苗预防，SARS 整个流行仅持续 8～9 个月；且爆发和消失的时间点均突然；亦属史无前例。

4. SARS 和其他人冠状病毒疾病的流行病学和临床特点相去甚远，即不符合其他人类冠状病毒疾病之规律。

5. SARS 与 SARS-CoV 和近年来出现的新发传染病之病毒进化和流行过程之普遍规律反常。

Final Dialogues

23rd December 2021

I told my wife that I had completed the manuscript. She smiled and hugged me tightly.

I told her, "Am I doing the right thing? I am a little afraid inside."

"Don't worry, and you are doing the right thing." She said.

"I have just attempted my best," I said.

"Do not worry, and your attempts will not be wasted. All the hard work done in late nights and early mornings will pay off," She said.

I also called Darpan and informed him that the manuscript was finally completed. He was very happy and, as usual, started questioning.

"So, what is the conclusion?" he asked

"All the circumstantial evidence is only in favor of the laboratory-made origin of SARS-CoV-2. However, even SARS-CoV was also strongly considered man-made by Xu Dezhong, Professor, Department of Military Epidemiology, Fourth Military Medical University; INCLEN, CEU, Director. China," I answered.

"Ok, so what about the natural origin of SARS-CoV-2?" he asked.

"Even at the time of SARS-CoV pandemic in 2002-2004, the outbreak pattern was suspicious, and I have already translated important fragments of that book. Not only one time but repetitively one term is quoted, "Unnatural Origin of SARS," that too with the scientific data and the pattern of SARS outbreak." I replied to Darpan

"And what about the bioweapons?" He asked.

"Yes, as I said, the officials of Department of Military Epidemiology, Fourth Military Medical University have extensively talked about bioweapons. They included various types of bioweapons and parts. For example, according to Xu Dezhong, the SARS-CoV is a man-made virus, and through it, the SARS pandemic occurred in the years 2002-2003-2004. But as he writes in the book, there was a big difference in the epidemic from 2002-2003 and 2004. This reason is enough to conclude that SARS-CoV-2 is unnatural and Laboratory originated. Because if SARS-CoV had no natural ancestor and similarly SARS-CoV-2 also have no natural ancestor. As Xu Dehzong writes in this book, 'there are SARS-like coronaviruses are present but not SARS-CoV itself. Therefore, there is a difference in SARS- Like the corona virus and SARS Coronavirus. So, SARS-CoV could not be considered as naturally originated.' So if SARS-CoV could not be considered as naturally originated, then we could not consider SARS-CoV-2 as naturally originated. It is also clear in the genetic studies of SARS-CoV-2." I completed my point.

"Good, then what about earlier cases of Wuhan Seafood market?" he asked with curiosity in his voice.

"You know that in case of SARS pandemic in 2002, the civet cats were found positive. Xu Dezhong and his team found it suspicious why the urban area civet cats were infected with SARS-CoV? Why not the rural and wild area civet cats not found infected with it? He and his team concluded that the civet cats were infected deliberately with lab. made virus and then brought into the Guangzhou seafood/ animal market. So, a similar case could have happened in the case of SARS-CoV-2." I replied.

"So, you want to say China could be the victim?" He asked.

"Maybe and maybe not. It could have been a part of the biological weapon's trial. But, on the other hand, it could have leaked or stolen from any laboratory-like Wuhan Institute of Virology, Wuhan CDC lab. or any military laboratory where research on SARS-CoV was in progress. And it is also possible that any animal could have deliberately infected to spread the virus among humans." I replied him.

"So, what is the conclusion?"

"I am not an expert of virology or epidemiology, but One thing is clear that the SARS-CoV-2 was made in the laboratory. As far as outbreak concern, we can not rule out any possibility from accidental leak to the trial of bioweapons." I replied.

"So, what is next?" he asked.

"This is not the end, and my analysis says it needs more insights and investigations to find what happened. I have some more analysis and conclusions, but they fall short due to evidences. But I am sure I or someone will find them too sooner or later. I will give my full efforts to find them. Still, we can not rule out the possibility of the Natural origin of SARS-CoV-2 completely. Still, at the same time, all the genetic researches and circumstantial evidence indicate only the laboratory origin of SARS-CoV-2. The work of Xu Dezhong and team adds more weight in the Lab. leak possibility." I replied

"You talked about some critical analysis can we expect PART-2 of this book? I am happy for you; best wishes and see you soon!" he replied.

"I have not decided yet for PART-2. Thank you, my friend, for everything, and see you soon!" I replied and disconnected the call.

Notes And References

Chapter: 3 The Dark history

- Information of Alan Schnur https://ccnmtl.columbia.edu/projects/caseconsortium/casestudies/112/casestudy/www/layout/case_id_112_id_226_c_bio.html
- Bat cave solves the mystery of deadly SARS virus — and suggests new outbreak could occur https://www.nature.com/articles/d41586-017-07766-9

- RaTG13 and its relativity with SARS-CoV-2https://www.preprints.org/manuscript/202005.0322/v2
- Inside the SARS outbreak: https://www.youtube.com/watch?v=0n7pIVS3t08

Chapter: 4 The Epicenter

- Sharry Markson's Interview. https://youtu.be/oh2Sj_QpZOA
- Critical Review Edward Holmes: https://zenodo.org/record/5112546#.YaeKBNBBzIU
- WHO mistake with the reading map. : https://zenodo.org/record/5202258#.YaeWCtBBzIU
- Did a Review of Samples Collected from a Mineshaft Cause the COVID-19 Pandemic?: https://zenodo.org/

record/4063813#.Yansw9BBzIU

- South China Morning Post report on Wuhan Seafood Market as the origin of Pandemic: https://www.youtube.com/watch?v=kX0knxioVwA
- Did the SARS-CoV-2 virus arise from a bat coronavirus research program in a Chinese laboratory? Very possibly. By Milton Leitenberg, June 4, 2020: https://archive.md/5m1N0#selection-999.0-999.302
- Wuhan Institute of Virology, CAS, “Take a look at the largest virus bank in Asia,” 2018, http://english.whiov.cas.cn/ne/201806/t20180604_193863.html
- Jon Cohen and Kai Kupferschmidt, “NIH-halted study unveil ist massive analysis of bat coronaviruses,” Science, June 2, 2020,: https://www.sciencemag.org/news/2020/06/nih-halted-study-unveils-its-massive-analysis-bat-coronaviruses

- Agence France Presse, “Wuhan lab had three live bat coronaviruses: Chinese state media,” May 24, 2020, https://au.news.yahoo.com/wuhan-lab-had-three-live-bat-coronaviruses-chinese-033127925–spt.html
- John Xie, “Chinese Lab with Checkered Safety Record Draws Scrutiny over Covid-19,” VOA News, April 21, 2020, https://www.voanews.com/covid-19-pandemic/chinese-lab-checkered-safety-record-draws-scrutiny-over-covid-19
- “流行性感冒诊疗方案（2019年版）.” National Health Commission of the People’s Republic of China, 2019. https://archive.is/zup74
- The Author. “How Many COVID-19 Cases Were Misreported as Influenza in China?”. Telegram, 2020. https://graph.org/2019-Influenza-Plan-08-29

- "2019版流感方案公布　新添医院感染控制措施." Sina Weibo News, 14 November 2019. https://archive.is/xOcDH
- National Health Commission of the People's Republic of China. "医务人员流感培训手册（2019　年版）." Government of Shanghai, 2019. https://archive.is/XrRUV
- "【重磅】全国流感医疗救治专家组专家名单公布！" Health Daily News, 21 October 2019. https://archive.is/fLpkx
- Jian, Jing Chu. "流感季防控处方：全民接种"社会疫苗." KK News, 20 December 2019. http://archive.is/hEXwH
- "国家卫生健康委举行新闻发布会 介绍我国流感防控工作有关情况." Government of the People's Republic of China, 30 October 2019. https://archive.is/Ze0Ww
- "流感进入冬春季活跃期　湖北疫情可防可控可治." Government of Hubei, 22 December 2019. http://archive.is/pLDCv
- Influenza Cases 2017-2019 Sino insider https://sinoinsider.com/2020/06/risk-watch-atypical-influenza-data-hints-at-earlier-transmission-of-coronavirus-in-china/
- Bats were never part of food source in Wuhan: https://img-prod.tgcom24.mediaset.it/images/2020/02/16/114720192-5eb8307f-017c-4075-a697-348628da0204.pdf
- Wilderness Youth | The Invisible Line of Defense (English Subtitled Version): https://www.youtube.com/watch?v=ovnUyTRMERI
- Wuhan And Death Toll: https://sinoinsider.com/2020/01/risk-watch-wuhan-coronavirus-death-count-could-be-35-times-higher-than-the-official-figure/

- SARS-CoV and Zoonosis Relativity: https://doi.org/10.1128/JVI.80.9.4211-4219.2006
- Wang YanYi Interview! Deleted Youtube Interview Video link: https://www.youtube.com/watch?v=bRuzsPA4Ukw, Video of that interview: https://twitter.com/cgtnofficial/status/1264220555013570562?lang=it
- Exclusive Eyewitness Account: I Saw Over 1000 Bodies in Zhongnan Hospital of Wuhan University: https://www.youtube.com/watch?v=aGOWwOTxLeE

Chapter: 5 The Ghost Cousin

- A pneumonia outbreak associated with a new coronavirus of probable bat origin: https://www.nature.com/articles/s41586-020-2012-7
- Wang YanYi Interview! Deleted Youtube Interview Video link: https://www.youtube.com/watch?v=bRuzsPA4Ukw Video of that interview: https://twitter.com/cgtnofficial/status/1264220555013570562?lang=it
- Origin and evolution of pathogenic coronaviruses by Shi ZhengLi: https://www.nature.com/articles/s41579-018-0118-9
- Why misinformation about COVID-19's origins keeps going viral: https://www.nationalgeographic.com/science/article/coronavirus-origins-misinformation-yan-report-fact-check-cvd

- Li-Meng Yan Zenodo Paper link : https://zenodo.org/

record/4028830#.YaUZXdBBzIU

- Major Concerns on the Identification of Bat Coronavirus Strain RaTG13 and Quality of Related Nature Paper: https://www.preprints.org/manuscript/202006.0044/v1

Chapter: 6 The Experiment

- A SARS-like cluster of circulating bat coronaviruses shows potential for human emergence by Vineet D Menachery, Boyd L Yount Jr, Kari Debbink, Sudhakar Agnihothram, Lisa E Gralinski, Jessica A Plante, Rachel L Graham, Trevor Scobey, Xing-Yi Ge, Eric F Donaldson, Scott H Randell, Antonio Lanzavecchia, Wayne A Marasco, Zhengli-Li Shi & Ralph S Baric: https://www.nature.com/articles/nm.3985
- What is a gain of function experiment: https://www.ncbi.nlm.nih.gov/pmc/articles/PMC4996883/
- Gain of Function in Brief: https://en.wikipedia.org/wiki/Gain-of-function_research
- SARS-CoV-2: the "Uncensored" Truth about Its Origin and Adipose-Derived Mesenchymal Stem Cells as New Potential Immune-Modulatory Weapon by Pietro Gentile: https://pubmed.ncbi.nlm.nih.gov/33815867/
- Did Corona Virus came From the lab. by By Milton Leitenberg, June 4, 2020: https://thebulletin.org/2020/06/did-the-sars-cov-2-virus-arise-from-a-bat-coronavirus-research-program-in-a-chinese-laboratory-very-possibly/
- Prevalence of IgG Antibody to SARS-Associated Coronavirus in Animal Traders --- Guangdong Province,

China, 2003 https://www.cdc.gov/mmwr/preview/mmwrhtml/mm5241a2.htm

- Bat Severe Acute Respiratory Syndrome - Like Coronavirus WIV1 Encodes an Extra Accessory Protein, ORFX, Involved in Modulation of the Host Immune Response by Lei-Ping Zenga, Yu-Tao Gaoa, Xing-Yi Gea, Qian Zhanga, Cheng Penga, Xing-Lou Yanga, Bing Tana, Jing Chena, Aleksei A. Chmurahttps://orcid.org/0000-0001-5544-0431b, Peter Daszakb, and Zheng-Li: https://journals.asm.org/doi/10.1128/JVI.03079-15
- State Department cables warned of safety issues at Wuhan lab studying bat coronaviruses: https://www.washingtonpost.com/opinions/2020/04/14/state-department-cables-warned-safety-issues-wuhan-lab-studying-bat-coronaviruses/
- BSL-4 was fully operational in 2018 : https://www.youtube.com/watch?v=GVQSzSVp5Jo
- Gravitas: Caught On Camera: Bats locked up inside the Wuhan lab: https://www.youtube.com/watch?v=LFdGkAFvW2U&t=184s

Chapter: 7 The Killer Couple

- The role of the furin cleavage site in SARS-CoV-2 spike protein-mediated membrane fusion in the presence or absence of trypsin by Shuai Xia, Qiaoshuai Lan, Shan Su, Xinling Wang, Wei Xu, Zezhong Liu, Yun Zhu, Qian Wang, Lu Lu & Shibo Jiang: https://www.nature.com/articles/s41392-020-0184-0
- SARS-CoV-2 a dagger to the aging heart by By Sally Robertson, B.Sc. : https://www.news-medical.net/news/

20200709/SARS-CoV-2-a-dagger-to-the-aging-heart.aspx

- Furin Cleavage Site Is Key to SARS-CoV-2 Pathogenesis: https://doi.org/10.1101/2020.08.26.268854
- SARS-CoV-2 spike and its adaptable furin cleavage site Gary R Whittaker: https://www.thelancet.com/journals/lanmic/article/PIIS2666-5247(21)00174-9/fulltext
- An open debate on SARS-CoV-2's proximal origin is long overdue by Rossana Segreto1*, Yuri Deigin2, Kevin McCairn3, Alejandro Sousa4,5, Dan Sirotkin6, Karl Sirotkin6, Jonathan J. Couey7, Adrian Jones8, Daoyu Zhang9: https://arxiv.org/ftp/arxiv/papers/2102/2102.03910.pdf
- The genetic structure of SARS-CoV-2 does not rule out a laboratory origin by Rossana Segreto, Yuri Deigin: https://onlinelibrary.wiley.com/doi/10.1002/bies.202000240

Chapter: 8 The Invisible Weapon

- SARS-CoV-2: the "Uncensored" Truth about Its Origin and Adipose-Derived Mesenchymal Stem Cells as New Potential Immune-Modulatory Weapon by PietroGentile: https://www.ncbi.nlm.nih.gov/pmc/articles/PMC7990360/#b5-ad-12-2-330
- Amazon Book Link of book, ' The Unnatural Origin of SARS and New Species of Man-Made Viruses as Genetic Bioweapons' by Xu Dehzong and Feng Li (Chinese): https://www.amazon.com/dp/B012984EOI
- Quotes: https://www.azquotes.com/author/16154-Mao_Zedong
- Chinese military scientists held talks on bio-weapon

benefits by SHARRI MARKSON, LIAM MENDES, JACK HAZLEWOOD: https://www.theaustralian.com.au/nation/politics/chinese-talks-on-biowar-benefits/news-story/977df9f170ec0f0d8e502a01ddd81554

- Unit 731: The horrors of biological warfare experiments that the world forgot By Sanghmitra: https://www.opindia.com/2021/06/unit-731-japan-china-usa-russia-biological-warfare-bioweapons-plagues-cholera-world-war/
- The spike protein of SARS-CoV — a target for vaccine and therapeutic development by
- Lanying Du,1 Yuxian He,1 Yusen Zhou,2 Shuwen Liu,3 Bo-Jian Zheng, and Shibo Jiang: https://www.ncbi.nlm.nih.gov/pmc/articles/PMC2750777/
- Serial Passage the Classical rout of Gain-of-Function Research. : https://twitter.com/ZionaEmanuel/status/1392365785188904964?s=20

Chapter: 9 From the Magician's Hat

- The Unnatural Origin of SARS and New Species of Man-Made Viruses as Genetic Bioweapons by Xu Dezhong and Li Feng was obtained from the Internet.

Satyamev Jayate
Jay Hind

www.ingramcontent.com/pod-product-compliance
Ingram Content Group UK Ltd.
Pitfield, Milton Keynes, MK11 3LW, UK
UKHW041955190726
13854UKWH00005B/1980

9 798885 464314